International Conference on Artificial Intelligence Applications and Technologies (AIAAT 2017)

IOP Conference Series: Materials Science and Engineering
Volume 261

Hawaii, USA
30 August - 2 September 2017

Editors:

Dan Zhang
Du Ruxu

ISBN: 978-1-5108-5114-6
ISSN: 1757-8981

Printed from e-media with permission by:

Curran Associates, Inc.
57 Morehouse Lane
Red Hook, NY 12571

proceedings.com

Some format issues inherent in the e-media version may also appear in this print version.

Printed by Curran Associates, Inc. (2017)

For permission requests, please contact the Institute of Physics
at the address below.

Institute of Physics
Dirac House, Temple Back
Bristol BS1 6BE UK

Phone: 44 1 17 929 7481
Fax: 44 1 17 920 0979

techtracking@iop.org

Additional copies of this publication are available from:

Curran Associates, Inc.
57 Morehouse Lane
Red Hook, NY 12571 USA
Phone: 845-758-0400
Fax: 845-758-2633
Email: curran@proceedings.com
Web: www.proceedings.com

Table of contents

Volume 261

2017 International Conference on Artificial Intelligence Applications and Technologies (AIAAT 2017)

30 August to 2 September 2017, Hawaii, USA

Accepted papers received: 24 October 2017
Published online: 6 November 2017

Preface

Machine Learning

Signal Processing and Control

Pattern Recognition

IOP Conf. Series: Materials Science and Engineering **261** (2017) 011001 doi:10.1088/1757-899X/261/1/011001

PREFACE

AIAAT 2017 was held in Hawaii, USA during Aug.30-Sept.2, 2017. AIAAT 2017 was organized by Hong Kong Society of Mechanical Engineers (HKSME). The conference provides a useful and wide platform both for display the latest research and for exchange of research results and thoughts in Artificial Intelligence Applications and Technologies. The participants of the conference were from almost every part of the world, with background of either academia or industry, even well-known enterprise. The success and prosperity of the conference is reflected high level of the papers received.

The proceedings are a compilation of the accepted papers and represent an interesting outcome of the conference. This book covers 3 chapters: Machine Learning; Signal Processing and Control; Pattern Recognition.

We would like to acknowledge all of those who supported AIAAT 2017. Each individual and institutional help were very important for the success of this conference. Especially we would like to thank the organizing committee for their valuable advices in the organization and helpful peer review of the papers.

We sincerely hope that AIAAT 2017 will be a forum for excellent discussions that will put forward new ideas and promote collaborative researches. We are sure that the proceedings will serve as an important research source of references and the knowledge, which will lead to not only scientific and engineering progress but also other new products and processes.

Prof. Dan Zhang, York University, Canada
Prof. Du Ruxu, The Chinese University of Hong Kong, HK

IOP Conf. Series: Materials Science and Engineering **261** (2017) 011002 doi:10.1088/1757-899X/261/1/011002

COMMITTEES

Conference Chairman

Prof. Dan Zhang

Professor and Kaneff Research Chair in Advanced Robotics and Mechatronics

Department of Mechanical Engineering, York University, Canada

EIC, ASME and CSME Fellow; IEEE and SME senior member

Conference Committee Chair

Prof. Du Ruxu, The Chinese University of Hong Kong, Hong Kong

Program Committee Chair

Prof. Ruili Wang, Massey University, New Zealand

Publication Chair

Prof. Wilson Q. Wang, Lakehead University, Canada

Technical Program Committee

Dr. Martin Stommel, AUT University, Auckland, New Zealand

Prof. Krzysztof Jemielniak, Warsaw University of Technology, Poland

Prof. Qin Jiahu, University of Science and Technology of China, China

Prof. Shao Cheng, Dalian University of Technology, China

Dr. Tansel Yucelen, University of South Florida, USA

Dr. Irene Fassi, Institute of Industrial Technologies and Automation, National Research Council, Italy

Prof. Xiao-Zheng Jin, HeFei University of Technology, China

Dr. Bin Wei, York University, Canada

Peer review statement

All papers published in this volume of *IOP Conference Series: Materials Science and Engineering* have been peer reviewed through processes administered by the proceedings Editors. Reviews were conducted by expert referees to the professional and scientific standards expected of a proceedings journal published by IOP Publishing.

IOP Conf. Series: Materials Science and Engineering **261** (2017) 012001 doi:10.1088/1757-899X/261/1/012001

Learning from Demonstration: Generalization via Task Segmentation

N Ettehadi[1], S Manaffam[1] and A Behal[1]

[1]Authors are with ECE Department and and NanoScience Technology Center, University of Central Florida (UCF), Orlando, FL 32816, USA

E-mail: nabil.ettehadi@knights.ucf.edu

Abstract. In this paper, a motion segmentation algorithm design is presented with the goal of segmenting a learned trajectory from demonstration such that each segment is locally maximally different from its neighbors. This segmentation is then exploited to appropriately scale (dilate/squeeze and/or rotate) a nominal trajectory learned from a few demonstrations on a fixed experimental setup such that it is applicable to different experimental settings without expanding the dataset and/or retraining the robot. The algorithm is computationally efficient in the sense that it allows facile transition between different environments. Experimental results using the Baxter robotic platform showcase the ability of the algorithm to accurately transfer a feeding task.

1. Introduction

One of the key challenges of learning from demonstration is to generalize a learned task to different situations and environments. This generalization to a new environment should be done such that the key features and constraints of the task (*i.e.*, key motions required to accomplish the task) be satisfied resulting in the successful implementation of the task in the new environment. In [1], a programming-by-demonstration framework is presented in order to generically extract a trajectory having the relevant features of a given task. The authors in [1] use Gaussian Mixture Model/Regression (GMM/GMR) to model the task and generate new trajectories. The GMM/GMR approach has been also used in [2] for manipulation tasks and is robust to the movement of obstacles included in the demonstrations. A number of methods have previously attempted to generalize a learned trajectory to different environments. The GMM/GMR is modified in [3] to adapt the means and covariances of a GMM to different locations and orientations of objects involved in a task. In [4], [5], and [6] the authors present a demonstration-guided motion planning (DGMP) to generate plans that avoid obstacles in a new environment while maintaining the critical features of the task. Another work has attempted to find sub-optimal controllers which reproduce a set of constraints of a given task [7]. However, the constraints are selected from a set of predefined constraints, *e.g.*, certain orientations for the end-effector. In [8], the authors use an invariant trajectory representation to separate essential motion information from context-specific information of the recorded demonstrations. Another work [9] uses dynamic movement primitives that are learned from human demonstrations with generalization to different start and goal points for a task.

In this paper, we propose an alternative approach to generalize a learned trajectory of a given task to different environments. In majority of the approaches discussed above, in order to learn and execute a task, a relatively big dataset, spanning the space in which the task would be performed, is needed.

However, our method does not require a big dataset; instead, a small number of demonstrations is needed to collect data with a robot for a fixed setup of the environment (no change in the position/orientation of the objects involved in the task). Consequently, the amount of data needed for training greatly reduces. Our approach takes the learned trajectory of the task for a fixed environment as an input and outputs an appropriate trajectory which successfully implements the learned task in a different environment. The approach consists of two parts: 1) segmentation of the learned trajectory w.r.t different motions that take place in each part of the trajectory, followed by 2) proper rotation and scaling of each segment based on the changes between the new environment and the environment in which the robot has been trained. The learned trajectory (input) is derived using the approach in [1]. Unlike neural network based methods, whose application is limited only to interpolation between the space over which they have been trained, our method is capable of generalization of the task to different environments even outside the training environment without the need for retraining the robot. Our approach only scales and rotates different segments of the trajectory, therefore it has less computational complexity than the approaches in [4]–[6]. Our method has been validated by running a task of feeding cereal from a bowl to a user.

2. Preliminaries

2.1. Notations and Definitions

The set of real n-vectors is denoted by $\mathbb{R}^n$ and the set of real $m \times n$ matrices is denoted by $\mathbb{R}^{m\times n}$. Matrices and vectors are denoted by capital and lower-case bold letters, respectively. The set of natural numbers is referred to by $\mathbb{N}$. Uppercase letters are used for constants and lowercase letters for scalar variables. The end-effector's trajectory is denoted by $\mathbf{X}(t) = [\mathbf{X}_{pos}(t); \mathbf{X}_{ori}(t)] \in \mathbb{R}^{6\times N}$ where $\mathbf{X}_{pos}(t) = [\mathbf{x}(t); \mathbf{y}(t); \mathbf{z}(t)] \in \mathbb{R}^{3\times N}$ is a matrix containing 3 time-dependent vectors for Cartesian coordinates (positions), $\mathbf{X}_{ori}(t) = [\theta_x(t); \theta_y(t); \theta_z(t)] \in \mathbb{R}^{3\times N}$ is a matrix containing 3 time-dependent vectors for the orientation of the end-effector (*e.g.,* Euler angles), and N is the length of the trajectory (number of time samples). $\mathbf{R}_{\beta,\mathbf{a}} = [\mathbf{c}_1\ \mathbf{c}_2\ \mathbf{c}_3] \in \mathbb{R}^{3\times 3}$ represents the rotation matrix around an arbitrary axis, $\mathbf{a}$, by an angle, β, where $\mathbf{c}_k \in \mathbb{R}^{3\times 1}$ represents the kth column vector of the rotation matrix $\mathbf{R}_{\beta,\mathbf{a}}$.

2.1.1. Definition *(Landmark)* Any entity relevant to the task influencing the behavior of the robot while the task is being performed. It is assumed that the shape of all landmarks do not change over the task. Landmark i is denoted by l_i.

2.1.2. Definition *(Nominal Setup for Landmarks)* Given a certain task, we define, as nominal, a known fixed setup in which the robot has been trained to perform that task. Specifically, the nominal setup, denoted by s_{nom}, is comprised of the fixed positions and/or orientations of all landmarks related to the task.

2.1.3. Definition *(Nominal Trajectory)* A given task is modelled as a Gaussian Mixture Model (GMM) using data collected from demonstrations (via kinesthetic teaching) performing the task in s_{nom}. The nominal trajectory of a given task, denoted by $\mathbf{X}_{nom}(t) = [\mathbf{X}_{nom_{pos}}(t); \mathbf{X}_{nom_{ori}}(t)] \in \mathbb{R}^{6\times N}$, is the estimated mean of the GMM, containing all of the essential relevant features (key motions) of the task.

2.2. Motivation and Problem Definition

One can obtain a nominal trajectory for a given task by using the methodology described in [1]. One interesting question is how to generalize the nominal trajectory to a new setup different from the nominal setup (outside the training set) without expanding the training set for retraining the robot. While most Machine Learning/Neural Networks techniques are capable of interpolation, they are not able to produce a proper trajectory for the sets outside the subspace spanned by the training samples [10]. Hence, we need other solutions to the following problem: *Let* $\mathbf{X}_{nom}(t) \in \mathbb{R}^{6\times N}$ *be the nominal trajectory of a given task,* T*, performed in the nominal setup,* s_{nom}*. Given a new setup* s *with the same landmarks, find*

$\mathbf{X}_{new}(t) \in \mathbb{R}^{6\times N}$ *performing* T *in* s , using $\mathbf{X}_{nom}(t)$, s_{nom} *and* s, *without retraining the robot?*
Please note that in the problem above, there are no assumptions imposed on the distances and the positions between the landmarks in the new setup s compared to s_{nom}. Let us consider a trajectory $\mathbf{X}_{nom}(t)$ of an arbitrary given task T, with n landmarks in the nominal setup s_{nom}. Now consider a new setup s. At first glance, the solution to the above problem is to rotate and scale each part of $\mathbf{X}_{nom}(t)$ lying between two consecutive landmarks l_i and l_{i+1}, according to their rotation and scaling between s_{nom} and s. However, by using this method the key motions of $\mathbf{X}_{nom}(t)$ would not be preserved. For example, consider a feeding task with two landmarks: a bowl full of cereal and a human user (the spoon is strictly attached to the end-effector). Scooping cereal is a key motion having a different nature compared to other key motions of this task (*e.g.,* keeping the loaded spoon level). Using the technique in [1], we can find $\mathbf{X}_{nom}(t)$ of this task in s_{nom}. Now, let us move the bowl away from the user defining the new setup s. As discussed above, by uniform stretching of the entire trajectory between the only existing landmarks, the robot would continue scooping even outside of the bowl since the corresponding segment of the trajectory within the bowl is also stretched; this would result in the end-effector tipping the bowl over. Hence, uniform scaling and/or rotation of a nominal trajectory (in general) does not guarantee the successful implementation of a given task. To avoid this problem, a segmentation technique is needed to divide the nominal trajectory of a given task into different segments. This segmentation technique should distinguish the different key motions taking place at each part of the task and segment the nominal trajectory accordingly. By having such segments, the nominal trajectory can be scaled and rotated intelligently (*i.e.,* segment by segment) during execution according to the differences between the two setups s and s_{nom}. Hence, we can state our problem definition as follows: *Let* $\mathbf{X}_{nom}(t) = [\mathbf{X}_{nom_{pos}}(t); \mathbf{X}_{nom_{ori}}(t)] \in \mathbb{R}^{6\times N}$ *be the nominal trajectory of a certain task,* T*, in the nominal setup,* s_{nom}.

A. *Find all the segments of* $\mathbf{X}_{nom}(t)$ *corresponding to different key motions of the task* T.

B. *Now let the number of the segments from problem 1 be* M*. Given a new setup* s *with the same landmarks in different positions and/or orientations, find the set* $\mathcal{M} = \{\mathbf{R}_{\theta_i,\mathbf{a}_i}, \alpha_{c,i}, \forall i \in \{1, ..., M\}$ and $\forall c \in \{x, y, z\}\}$.

3. Problem A: Segmentation of the Nominal Trajectory

In the previous section, we divided the problem into two parts. The first part is intended to segment the nominal trajectory corresponding to different key motions of the given task. A motion is constructed by a set of velocity commands (both linear and angular). At each segment, certain velocities, having specific directions and amplitudes, are commanded to perform the appropriate motion required for that part of the task. Thus, for the purpose of the segmentation, we base our analysis on the nominal velocity trajectory $\mathbf{\Lambda}(t) \in \mathbb{R}^{6\times N}$. In lieu of a binary representation as motion/no-motion, we embed the motion in a higher dimensional space by representing $\mathbf{\Lambda}(t)$ as a quantized level-valued signal. Next, we discuss quantization and segmentation.

3.1. Quantization

Nominal velocity trajectory $\mathbf{\Lambda}(t) \in \mathbb{R}^{6\times N}$ is quantized by the following approach: we uniformly quantize linear and angular velocity components into q_v and q_ω levels, respectively. The length of two consecutive levels in our quantization are denoted by δ_v and δ_ω for linear and angular velocity, respectively. Thus, the linear velocity signals belonging to interval $(-\delta_v, +\delta_v)$ and angular velocity signals belonging to $(-\delta_\omega, +\delta_\omega)$, are mapped to level 0 denoting no motion. The choice of quantization steps can be formulated as a detection problem where δ_v and δ_ω are calculated for a chosen small probability of false alarm[1]. For more details regarding this subject, the readers are referred to literature in detection theory [11]. The number of sets q_v and q_ω are derived as follows: $q_v = \max\{v_x(i), v_y(i), v_z(i)\}/\delta_v, q_\omega = \max\{\omega_x(i), \omega_y(i), \omega_z(i)\}/\delta_\omega$. Each quantization level is

[1] The probability that there is no motion taking place in reality but a motion is falsely detected.

labeled with d_{v_x}, d_{v_y}, and d_{v_z} corresponding to v_x, v_y and v_z, respectively (*e.g.*, $d_{v_x}(i) = 2$ means that v_x at time stamp i falls into the second level of our quantization). Similarly, we use $d_{\omega_x}, d_{\omega_y}$, and d_{ω_z} for the quantizations level labels of angular velocities. Now, we have a quantized nominal velocity trajectory of the end-effector denoted by $\mathbf{D}_{nom} = [\mathbf{d}_{nom}(i)]_{6\times N}$ for $i = 1, 2, ..., N$ where $\mathbf{d}_{nom}(i) = [d_{v_x}(i); d_{v_y}(i); d_{v_z}(i); d_{\omega_x}(i); d_{\omega_y}(i); d_{\omega_z}(i)] \in \mathbb{R}^6$ is the quantized nominal velocity vector of the end-effector at each time stamp i. It is important to mention that the quantization levels take into account both magnitude and sign of the velocities. This is required because two motions in opposite directions with the same magnitude should be distinguished from each other.

3.2. Segmentation Algorithm

$\mathbf{D}_{nom}$ is segmented by the following approach: we first uniformly divide $\mathbf{D}_{nom}$ into r mutually-exclusive, equally-lengthed units in time called search-segments: $\mathbf{D}_{nom} = \{\mathbf{D}_{nom_1}, \mathbf{D}_{nom_2}, ..., \mathbf{D}_{nom_r}\}, \mathbf{D}_{nom_i} \cap \mathbf{D}_{nom_j} = \varnothing, \bigcup_{i=1}^{r} \mathbf{D}_{nom_i} = \mathbf{D}_{nom}, \forall i \neq j \in \{1, 2, ..., r\}$. The number r is a parameter, chosen such that it is at least the size of the smallest acceptable task segment but significantly smaller than data size N; one good choice would be: $r \approx \frac{N}{10}$. The center of each search-segment is defined as the average of its members, *i.e.*, if search-segment $\mathbf{D}_{nom_i} \in \mathbb{R}^{6\times\lceil\frac{N}{r}\rceil}$ starts at time stamp f_i and ends at time stamp $f_i + \lceil\frac{N}{r}\rceil$, its center is calculated as: $\bar{\mathbf{d}}_{nom_i} = \sum_{j=f_i}^{f_i+\lceil\frac{N}{r}\rceil} \mathbf{d}_{nom_i}(j)/(\lceil\frac{N}{r}\rceil + 1)$. As previously stated, we aim to merge these search-segments to form the larger segments representing the different key motions taking place in different parts of the nominal trajectory. In order to do that, a measure of similarity between two consecutive search-segments should be defined to decide if two search-segments need to be merged or not. This measure is defined as the absolute value of the difference between the center of search-segment i and $i+1$: $\eta(i) = \left|\bar{\mathbf{d}}_{nom_{i+1}} - \bar{\mathbf{d}}_{nomi}\right|$, $\forall i = 1, 2, ..., r-1$. To reduce the effect of noise, for any two consecutive search-segments $i-1$ and i, the mean of the defined measure of time stamp $i-1$, *i.e.*, $\bar{\eta}(i-1)$, is compared with $\eta(i)$: if $\bar{\eta}(i-1) \leq \eta(i)$, it means there exists more similarities between the corresponding velocities in search-segment i and the mean of the members of the very last segment of this step, compared to the similarities between the corresponding velocities in search-segment i and $i+1$. Thus, search-segment i merges with the very last segment of this step; otherwise, the very last segment of this step is closed as a complete segment and the search for a new segment initializes with search-segment i being the first member of it. The following pseudo code provides the complete algorithm which returns seg_final as the output:.

Step 1. $\eta[r] \leftarrow \infty$, $k \leftarrow 1$, $\mathbf{D}_{nom_1} \leftarrow seg\_temp$, $\bar{\eta}[1] \leftarrow \eta[1]$ //initialization
Step 2. For $i \leftarrow 2$ to r do
  Step 2.1. If $\eta[i] \geq \bar{\eta}[i-1]$
    Step 2.1.1. $seg\_temp \leftarrow [seg\_temp, \mathbf{D}_{nom_i}[:, f_i : f_i + \lceil\frac{N}{r}\rceil]]$
    Step 2.1.2. $\bar{\eta}[i] \leftarrow \frac{(i-1)\times\bar{\eta}[i-1]+\eta[i]}{i}$
  Step 2.2. else
    Step 2.2.1. $seg\_final[k] \leftarrow seg\_temp$ //close seg_temp as a completed segment.
    Step 2.2.2 $k \leftarrow k+1$
    Step 2.2.3. $seg\_temp \leftarrow [\mathbf{D}_{nom_i}[:, f_i : f_i + \lceil\frac{N}{r}\rceil]]$
    Step 2.2.4. $\bar{\eta}[i] \leftarrow \eta[i]$

4. Problem B: Off-line Adaptation of The Nominal Trajectory to a New Setup for Landmarks

Now, by having the segmented nominal trajectory, we can properly scale and rotate each segment based on the differences between s and s_{nom}. Since the nominal trajectory needs to be rotated, here we use the rotation matrix representation for $\mathbf{X}_{nom_{ori}}(t)$ rather than Euler angle representation, *i.e.*, at each time stamp t we have $\hat{\mathbf{X}}_{nom_{ori}}(t) = [\mathbf{c}_1(t)\ \mathbf{c}_2(t)\ \mathbf{c}_3(t)] \in \mathbb{R}^{3\times3}$ denoting the rotation matrix for the orientation of the end-effector. Now, suppose a segment k starts at time index j and finishes at time index $j+\lceil N/r \rceil$. The new segment in $\mathbf{X}_{new}(t)$ is derived from the corresponding segment in $\tilde{\mathbf{X}}_{nom}$ as follows:

$$\mathbf{X}_{new_{pos}}(t) = \mathbf{X}_{new_{pos}}(t-1) + (\mathbf{A}_k \times \mathbf{R}_{\theta_k,\mathbf{a}_k} \times [\mathbf{X}_{nom_{pos}}(t) - \mathbf{X}_{nom_{pos}}(t-1)]) \quad (1)$$
$$\hat{\mathbf{X}}_{new_{ori}}(t) = \mathbf{R}_{\theta_k,\mathbf{a}_k} \times \hat{\mathbf{X}}_{nom_{ori}}(t) for\ t = j, j+1, j+2, ..., j+\lceil N/r \rceil - 1$$

where $\mathbf{A}_k = diag(\alpha_{x,k}, \alpha_{y,k}, \alpha_{z,k})$ is the scale factor matrix calculated using the corresponding scale factors in each Cartesian components and t is the time index. The initial condition for this equation, *i.e.*, $\mathbf{X}_{new}(j-1)$ is equal to the last point of the previous segment $k-1$, which has been previously properly rotated and scaled using (1). Note that since the shape of landmarks are assumed to be constant and both positions and orientations are tied with time stamps, it is sufficient to only scale the positions. By proper rotation and scaling of all segments, the characteristics of the motions in different segments (key motions taking place at different segments) will remain intact. Finally, by back projecting $\hat{\mathbf{X}}_{new_{ori}}(t)$ to Euler angle representation, *i.e.*, $\hat{\mathbf{X}}_{new_{ori}}(t) \rightarrow \mathbf{X}_{new_{ori}}(t)$, the appropriate trajectory of the task for setup s is achieved as: $\mathbf{X}_{new}(t) = [\mathbf{X}_{new_{pos}}(t); \mathbf{X}_{new_{ori}}(t)] \in \mathbb{R}^{6\times N}$.

5. Experimental Results

In order to demonstrate the validity of our method, we have chosen the task of feeding cereal to a human user. There are two landmarks involved in the task: landmark l_1 is a bowl full of cereal and l_2 is the mouth of a user which is considered to be a point in the space[2]. The nominal setup is $s_{nom} = \{x_{nom_b} = 0.806$, $y_{nom_b} = 0.811$, $z_{nom_b} = -0.070$, $x_{nom_m} = 0.858$, $y_{nom_m} = 0.489$, $z_{nom_m} = -0.001\}$meter w.r.t. the base frame of the robot. figure 1(a) shows the nominal setup for the task. We use a 7-DOF Baxter robot in order to conduct our experiment. An arbitrary coordinate frame is set to be placed at the center of the bowl and the positions and orientations of the spoon are being recorded w.r.t this coordinate frame. 11 demonstrations of the task are recorded by a human teacher performing the task in the fixed nominal setup s_{nom} with the robot using kinesthetic teaching. Using [1] the nominal trajectory is derived from the demonstrations. Next, we run our segmentation algorithm on $\mathbf{X}_{nom}(t)$ consisting of 870 samples. We test with two different number of initial search-segments, $r_1 = 28$ and $r_1 = 68$ which set the length of each search-segment to be $\lceil \frac{870}{28} \rceil = 32$, and $\lceil \frac{870}{68} \rceil = 13$, respectively. The algorithm results in 6 segments for $r_1 = 28$ and 7 segments for $r_2 = 68$ as shown in figures 1. As illustrated in figure 1(b), the algorithm is able to successfully capture total number of 6 segments; 3 segments for the bowl, namely, approaching the bowl (blue segment), scooping (red segment), and leaving the bowl (green segment), 1 segment for the translation (yellow) of the spoon from the bowl to mouth of the user, another segment for the reversed translation (black), and finally the last segment (light blue) as the resetting of the spoon in preparation for the next feeding cycle. For the case where $r_2 = 68$, we get 7 segments; however, the additional segment shown in figure 1(c) is a very small blue segment at the beginning of the trajectory likely resulting from the fact that increase in the number of search-segments may lead to artifacts related to reduction in window size. Nonetheless, the result that both $r_1 = 28$ and $r_2 = 68$ yield almost identical results for segmentation, shows that the algorithm is robust to the initial number of search-segments. Next, we discuss the results for experiments using the 3 different setups: $s_1 = \{0.457, 1.007, -0.07, 0.849, 0.704, -0.001\}$, $s_2 = \{0.564, 0.941, -0.07, 0.931, 0.773, -0.001\}$ and $s_3 = \{0.608, 0.896, -0.07, 0.761, 0.704, -0.001\}$ meter. Since the size of the bowl has not changed, as it was assumed in the definition of landmarks, only the parts of nominal trajectory corresponding to the first and the second translation phases (as described above) are scaled by their corresponding scale factor matrices $\mathbf{A}_1$ and $\mathbf{A}_2$, respectively. Moreover, the whole trajectory is rotated using $\mathbf{R}_{\theta,\mathbf{a}}$ (see equation (1)). For s_1 we have: $A_1 = diag[2.32, 2.29, 1]$, $A_2 = diag[2.32, 2.29, 1]$ and $\mathbf{R}_{43.15°,\mathbf{z}}$. For s_2 we have: $A_1 = diag[1.6, 1.6, 1]$, $A_2 = diag[1.23, 1.18, 1]$ and $\mathbf{R}_{56.26°,\mathbf{z}}$. For s_3 we have: $A_1 = diag[1.6, 1.6, 1]$,

[2] There is no specific action taken by the robot/user at the position of the mouth; only that the robot stops moving for an amount of time.

$A_2 = diag[0.75, 0.78, 1]$ and $\mathbf{R}_{29.40^\circ,\mathbf{z}}$. We successfully tested our algorithm on these 3 setups. The new trajectories for s_1, s_2, and s_3 are shown in figure 2. For better comparison, all the trajectories are plotted in their new bowl frames. As it can be seen, the segments corresponding to the bowl have remained intact while the other segments have been scaled properly to implement the task successfully for each setup.[3]

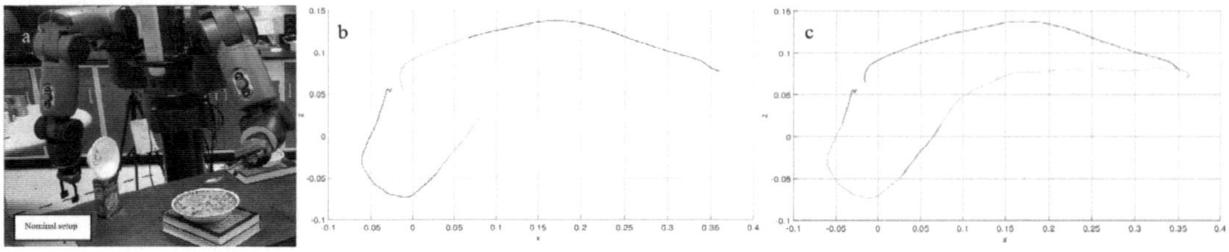

Figure 1. (a) Nominal setup for the task of feeding in our experiment. Projection of the segmented trajectory on zx plane with total number of search-segments (b) $r_1 = 28$ and (c) $r_2 = 68$

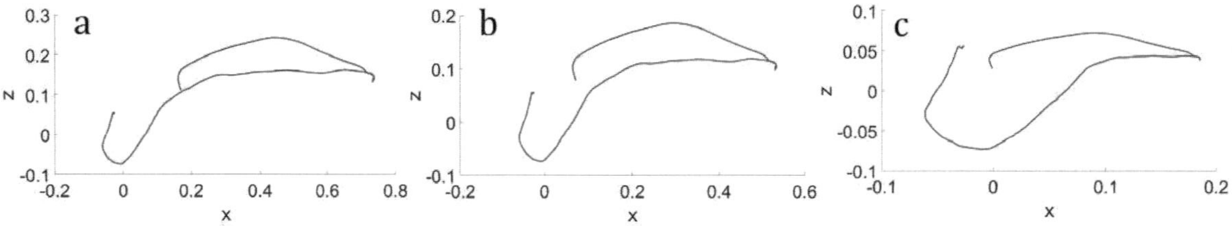

Figure 2. Scaled trajectories for setup (a) s_1, (b) s_2 and (c) s_3.

6. Conclusions

In this paper, we presented a method (a segmentation algorithm along with an off-line adaptation law) to generalize a learned nominal trajectory of a given task to different setups without expanding the training set to retrain the robot.The method was successfully validated on 3 different setups for a feeding task.

Acknowledgments:
This study was funded in part by NIDILRR grant # H133G120275, and in part by NSF grant numbers IIS-1409823 and IIS-1527794. However, these contents do not necessarily represent the policy of the aforementioned funding agencies, and you should not assume endorsement by the Federal Government.

References

[1] Calinon S, Guenter F and Billard A, "On learning, representing, and generalizing a task in a humanoid robot" April 2007. *IEEE Transactions on Systems, Man, and Cybernetics*, Part B (Cybernetics), vol. 37, no. 2, pp. 286-298, .

[2] Calinon S, et al. "Handling of multiple constraints and motion alternatives in a robot programming by demonstration framework" 2009. *Humanoid Robots, Humanoids 9th IEEE-RAS International Conference on* pp. 582-588.

[3] Calinon S, Alizadeh T and Caldwell G D. "On improving the extrapolation capability of task-parameterized movement models" 2013. *IEEE/RSJ International Conference on Intelligent Robots and Systems* pp. 610-616.

[4] Ye G and Alterovitz R, "Demonstration-guided motion planning" 2011. *International Symposium on Robotics Research.* vol. 5.

[3] The full video of the experiment is available at: https://youtu.be/3ecCBCUitqI

IOP Conf. Series: Materials Science and Engineering **261** (2017) 012001 doi:10.1088/1757-899X/261/1/012001

[5] Bowen C and Alterovitz R, "Closed-loop global motion planning for reactive execution of learned tasks," 2014. *IEEE/RSJ International Conference on Intelligent Robots and Systems, Chicago, IL*, pp. 1754-1760.
[6] Bowen C, Ye G and Alterovitz R. "Asymptotically optimal motion planning for learned tasks using time-dependent cost maps" 2015. *IEEE Transactions on Automation Science and Engineering 12.1*, pp. 171-182.
[7] Alissandrakis A, et al."An approach for programming robots by demonstration: Generalization across different initial configurations of manipulated objects" 2005. *International Symposium on Computational Intelligence in Robotics and Automation, IEEE*, pp. 61-66 .
[8] Vochten M, De Laet T and De Schutter J. "Generalizing demonstrated motions and adaptive motion generation using an invariant rigid body trajectory representation" 2016. *IEEE International Conference on Robotics and Automation (ICRA), Stockholm*, pp. 234-241.
[9] Paxton C, Hager G D and Bascetta L. "An incremental approach to learning generalizable robot tasks from human demonstration" 2015. *Robotics and Automation (ICRA), IEEE International Conference on*, pp. 5616-21
[10] Bishop C M. *Pattern Recognition and Machine Learning (Information Science and Statistics)* 2006. Springer-Verlag New York, Inc., Secaucus, NJ, USA.
[11] Wilmshurst T H. *Signal Recovery from Noise in Electronic Instrumentation*, Second Edition 1990. Taylor & Francis, ISBN: 978-0-7503-0058-2.

IOP Conf. Series: Materials Science and Engineering **261** (2017) 012002 doi:10.1088/1757-899X/261/1/012002

Some Tricks in Parameter Selection for Extreme Learning Machine

Weipeng Cao[1], Jinzhu Gao[2], Zhong Ming[1,*] and Shubin Cai[1]

[1] College of Computer Science and Software Engineering, Shenzhen University, Shenzhen 518060, China

[2] School of Engineering and Computer Science, University of the Pacific, CA 95211, USA

*Corresponding Author: mingz@szu.edu.cn

Abstract. Extreme learning machine (ELM) is a widely used neural network with random weights (NNRW), which has made great contributions to many fields. However, the relationship between the parameters and the performance of ELM has not been fully investigated yet, i.e. the impact of the number of hidden layer nodes, the randomization range of the weights between the input layer and hidden layer, the randomization range of the threshold of hidden nodes, and the type of activation functions. In this paper, eight benchmark functions are used to study this relationship. We have some interesting findings, such as more hidden layer nodes cannot guarantee the best performance of ELM, the empirical randomization range of the hidden weights (i.e., [-1, 1]) and the empirical randomization range of the threshold of hidden nodes (i.e., [0, 1]) may not lead to the optimal performance of ELM models, and ELM with *sigmoid* as the activation function always achieves better performance on some regression problems than ELM with *tribas* as the activation function. We hope the findings from our work could provide a useful guidance for researchers to select right parameters for ELM.

1. Introduction

Traditional artificial neural network (ANN), including deep learning, train models using the BP-based method [1]. That is, iteratively tuning all the weights and biases based on the back-propagated derivatives of loss function. This training mechanism has several notorious shortcomings such as local minima problem, slow convergence rate, time-consuming, and model uncertainty.

By contrast, neural network with random weights (NNRW) is a non-iterative algorithm that can avoid the above issues. NNRW was first proposed in 1992 by Schmidt et al. [2] and Pao's group [3], respectively. The biggest difference between NNRW and the BP-based method is that the hidden weights are selected randomly while the output weights are obtained analytically. Compared with the BP-based method, NNRW can achieve much faster training speed. The combination of NNRW and traditional deep learning can effectively improve the computational efficiency of deep learning, which has been proven in many applications [4-6].

The name of NNRW proposed by Pao's group [3] is random vector functional link networks (RVFL). In RVFL, the input layer is directly connected to both the hidden layer and the output layer. The hidden parameters are randomly selected from [-1, 1], while the weights between the input layer and output layer and the weights between the hidden layer and output layer are obtained by Moore–

Penrose pseudo-inverse. Zhang et al. [7] and Li's group [8] showed that the direct links between the input layer and the output layer have a significant impact on the performance of RVFL and the empirical randomization range (i.e., [-1, 1]) of input weights and biases may not always be the right one. In addition, they studied the impact of the different activation functions on the performance of RVFL.

The NNRW proposed by Schmidt et al. [2] is somewhat different from RVFL. In Schmidt's model, there is no direct link between the input layer and output layer and they used the Fisher method to determine the output weights. This type of NNRW was further investigated by Huang et al. who proposed a generalized algorithm, that is, extreme learning machine (ELM), in 2004 [9]. They have given a lot of theoretical analysis and rigorous theoretical proof for ELM [10, 11]. Huang et al. demonstrated that ELM has universal approximation capability and can approximate any continuous target function with probability one under certain conditions. In addition, any nonlinear piecewise continuous random hidden nodes can be used in ELM, such as Sigmoid nodes, Radial basis function nodes, Wavelet, etc. ELM and its variants have been widely used in real-world applications, including human face recognition [12, 13], fault diagnosis [14, 15], handwritten digit recognition [16], 3D image classification [17], fuzzy nonlinear regression [18], etc.

The idea behind RVFL, Schmidt's method, and ELM, is similar, but their implementations are different [19]. Inspired by [7, 8], we did a comprehensive study on the relationship between the parameters and the performance of ELM using eight different benchmark functions. The parameters we studied include empirical randomization range of hidden parameters, the number of hidden layer nodes, and the selection of activation functions. In this paper, we share some interesting findings that could provide a useful guidance to help researchers select parameters wisely.

The rest of this paper is organized as follows. Section 2 gives a brief introduction to ELM. The details of experiment settings are shown in Section3. In Section 4, we present the simulation results and our discussions. Conclusions and future work are given in Section 5.

2. Extreme learning machine

Extreme learning machine (ELM) is a special single hidden layer feed-forward neural network (SLFN), which was proposed by Huang's group in 2004 [9]. A typical structure of ELM is shown in Figure 1.

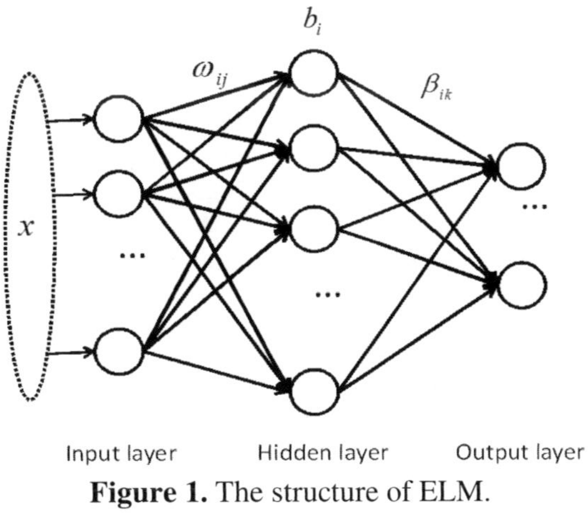

Figure 1. The structure of ELM.

where ω denotes the weights between the input layer and hidden layer, b denotes the threshold of hidden nodes, and β denotes the output weights. It is noted that there is an activation function in the hidden layer to perform nonlinear transformation.

The biggest difference between ELM and other traditional SLFNs (BP-based methods) is that the hidden parameters (the weights between the input layer and hidden layer and the threshold of hidden nodes) are randomly selected and the output weights are determined by solving a system of linear matrix equations. Compared with the BP-based methods, ELM can achieve much faster learning speed with acceptable accuracy.

A typical ELM has L hidden layer nodes, N instances, and an activation function $G(\chi)$, which can be modeled as:

$$\sum_{i=1}^{L} \beta_i G(\omega_{ij} \cdot x_j + b_i) = o_j, j = 1, 2, \ldots, N \tag{1}$$

where o_j is the predictive value of ELM model. According to the ELM theory [11], the optimization goal of ELM is

$$\min_{\|\beta\|} (\min \sum_{i=1}^{N} \| t_i - o_i \|^2), \tag{2}$$

where t is the actual value of each instance, and $\|\bullet\|$ is the matrix norm in Euclidean space. Equation (1) and (2) can be rewritten as

$$H\beta = T, \tag{3}$$

where $H = \begin{pmatrix} G(\omega_1 \cdot x_1 + b_1) & \cdots & G(\omega_L \cdot x_1 + b_L) \\ \vdots & \ddots & \vdots \\ G(\omega_1 \cdot x_N + b_1) & \cdots & G(\omega_L \cdot x_N + b_L) \end{pmatrix}_{N \times L}$, $\beta = \begin{bmatrix} \beta_1^T \\ \vdots \\ \beta_L^T \end{bmatrix}_{L \times m}$, $T = \begin{bmatrix} t_1^T \\ \vdots \\ t_N^T \end{bmatrix}_{N \times m}$

H is the output matrix of the hidden layer, and the output weight β can be solved by

$$\beta = H^{+} T \tag{4}$$

where H^{+} is the Moore–Penrose generalized inverse of H.

In general, the weights ω are generated randomly within the range of [-1, 1] and the thresholds of hidden layer nodes, b, are generated randomly within the range of [0, 1] under a uniform sampling distribution. The output weight β is determined by the Moore-Penrose generalized inverse as shown in equation (4). The commonly used activation functions $G(\chi)$ include *Sigmoid function*, *Sine function*, *Hardlim function*, *Triangular basis function*, and *Radial basis function*.

Although ELM and its variants have been widely applied to numerous applications, to the best of our knowledge, there is no thorough investigation on the relationship between the parameters and the performance of ELM. In this paper, we use eight benchmark functions to fully study the impact of these parameters on the performance of ELM by answering the following questions.

(1) Is it true: the more hidden layer nodes, the better performance of ELM?

(2) Can the empirical randomization range of hidden weights (i.e., [-1, 1]) lead to the optimal performance of ELM?

(3) Can the empirical randomization range of the hidden threshold (i.e., [0, 1]) lead to the optimal performance of ELM?

(4) What's impact of different activation functions on the performance of ELM?

3. Experiment settings

All our experiments are conducted in the MATLAB R2014a environment on the same Windows 10 machine with Intel Core i5-5300U 2.3 GHz CPU and 8 GB RAM. For each benchmark function, the average results over 50 trials are obtained for ELM.

3.1. Data preparation

In our experiments, eight commonly used benchmark functions are used to test the performance of ELM, i.e., *Sphere* function (Function 1), *Step* function (Function 2), *Schwefel* function (Function 3), *Rastrigin* function (Function 4), *Ackley* function (Function 5), *Rosenbrock* function (Function 6), *Schafferf* function (Function 7), and *Quartic* function (Function 8). The details of these eight functions are shown in Table 1.

Each benchmark function is used to generate a dataset with 2000 records, respectively. In each experiment, we shuffled the records, randomly selected 1000 records for training, and saved the remaining 1000 records for testing. The indicators for ELM performance testing include Testing Time, Standard Deviation of Testing Accuracy, the Root-Mean-Square Error of Training (Training RMSE), and the Root-Mean-Square Error of Testing (Testing RMSE).

Table 1. The details of eight benchmark functions

Function name	Function model	Initial range	$f_{\min}$
Sphere	$f=\sum_{i=1}^{D} x_i^2$	$[-100,100]^D$	0
Step	$f=\sum_{i=1}^{D} (\lfloor x_i+0.5 \rfloor)^2$	$[-100,100]^D$	0
Schwefel	$f=\sum_{i=1}^{D} x_i \times \sin\sqrt{\lVert x_i \rVert}$	$[-500,500]^D$	-12569.6
Rastrigin	$f=\sum_{i=1}^{D} \left[x_i^2-10\cos(2\pi x_i)+10 \right]$	$[-5.12,5.12]^D$	0
Ackley	$f=-20\exp(-0.2\sqrt{\frac{1}{D}\sum_{i=1}^{D} x_i^2})-\exp(\frac{1}{D}\sum_{i=1}^{D}\cos 2\pi x_i)+20+e$	$[-32,32]^D$	0
Rosenbrock	$f=\sum_{i=1}^{D-1} \left[100(x_{i+1}-x_i)^2+(x_i-1)^2 \right]$	$[-30,30]^D$	0
Schafferf	$f=\sum_{i=1}^{D-1} \left[(x_i^2+x_{i+1}^2)^{0.25}(\sin(50(x_i^2+x_{i+1}^2)^{0.1})^2+1) \right]$	$[-100,100]^D$	0
Quartic	$f=\sum_{i=1}^{D} ix_i^4+random[0,1)$	$[-1.28,1.28]^D$	0

3.2. Different ELM configurations
In our experiments, the variables for testing include the number of hidden layer nodes, the randomization range of the weights between the input layer and hidden layer, the randomization range of the threshold of hidden nodes, and the type of activation functions. In this paper, we use the single variable method to carry on the experiments, i.e., when we test the impact of a specific variable on the performance of ELM, the other variables remain unchanged. All the experimental results and analysis are shown in Section 4.

4. Simulation results and discussions
Supported by the experiments described in Section 3, we conducted a comprehensive study on the relationship between the parameters and performance of ELM using eight benchmark functions. The experimental results are shown in Table 2-5. It is noted that the near results are underlined and the best results are in boldface. In the rest of this section, we will give a detailed explanation on these results.

4.1. Is it true: the more hidden layer nodes, the better the performance of ELM?
Although Huang et al. have theoretically proved that the ELM model can approximate any continuous target function with any degree of accuracy provided it has as many hidden layer nodes as possible

[10]. However, in practical applications, would more hidden layer nodes lead to better performance of ELM model?

To answer this question, we conducted a validation experiment on the eight benchmark functions. The experimental results are shown in Table 2. In this experiment, the randomization range of the weights between the input layer and hidden layer is set to [-1, 1], the randomization range of the threshold of hidden nodes is set to [0, 1], and *Sigmoid function* is used as the activation function of ELM. Here, the variable of testing is the number of the hidden layer nodes L, which is set to

$$L=K*N, K=\{0.2, 0.4, 0.6, 0.8\} \quad (5)$$

where N is the number of training records.

Table 2. The performance of ELM based on different number of hidden layer nodes.

Dataset	Nodes number	Testing time(s)	Standard deviation	Mean	
				Training RMSE	Testing RMSE
Function1	0.2 * N	1.56E-02	4.98E-08	2.79E-08	3.62E-08
	0.4 * N	3.25E-02	1.83E-08	1.15E-08	**1.46E-08**
	0.6 * N	4.72E-02	**1.41E-08**	1.51E-08	1.71E-08
	0.8 * N	6.44E-02	8.88E-08	5.84E-08	5.95E-08
Function2	0.2 * N	1.59E-02	**6.80E-09**	6.18E-09	**7.33E-09**
	0.4 * N	3.16E-02	1.64E-07	3.92E-08	4.05E-08
	0.6 * N	4.41E-02	1.94E-07	8.14E-08	8.16E-08
	0.8 * N	6.22E-02	1.23E-07	7.38E-08	7.40E-08
Function3	0.2 * N	1.72E-02	1.67E-02	2.76E-01	**4.16E-01**
	0.4 * N	2.69E-02	**1.42E-02**	2.02E-01	4.84E-01
	0.6 * N	4.72E-02	2.17E-02	1.28E-01	5.85E-01
	0.8 * N	6.06E-02	3.92E-02	6.56E-02	7.46E-01
Function4	0.2 * N	1.75E-02	**2.92E-06**	1.93E-06	**1.93E-06**
	0.4 * N	3.03E-02	5.36E-06	4.90E-06	4.90E-06
	0.6 * N	4.63E-02	4.71E-06	5.20E-06	5.20E-06
	0.8 * N	6.34E-02	3.21E-06	3.70E-06	3.70E-06
Function5	0.2 * N	1.97E-02	2.79E-04	2.20E-04	3.07E-04
	0.4 * N	3.22E-02	1.42E-04	8.38E-05	1.52E-04
	0.6 * N	4.66E-02	**1.14E-04**	4.82E-05	1.35E-04
	0.8 * N	6.53E-02	1.24E-04	2.35E-05	**1.28E-04**
Function6	0.2 * N	1.69E-02	**5.77E-06**	6.75E-06	**5.77E-06**
	0.4 * N	3.13E-02	1.04E-05	1.13E-05	1.13E-05
	0.6 * N	4.66E-02	5.87E-06	7.63E-06	7.63E-06
	0.8 * N	6.19E-02	7.38E-06	1.03E-05	1.03E-05
Function7	0.2 * N	1.69E-02	**6.05E-08**	2.54E-08	**2.60E-08**
	0.4 * N	2.84E-02	7.00E-07	1.96E-07	1.96E-07
	0.6 * N	4.56E-02	1.99E-07	1.52E-07	1.52E-07
	0.8 * N	6.16E-02	7.45E-07	3.65E-07	3.65E-07
Function8	0.2 * N	1.50E-02	**1.61E-03**	7.86E-03	**1.93E-02**
	0.4 * N	3.34E-02	2.31E-03	2.86E-03	**1.93E-02**
	0.6 * N	5.19E-02	2.23E-03	9.51E-04	2.05E-02
	0.8 * N	6.56E-02	3.05E-03	3.02E-04	2.29E-02

From the Table 2, we can see that simply adding hidden layer nodes can not guarantee the better performance for ELM. In this experiment, the ELM model with 400 hidden layer nodes achieves the best performance on Function 1, while the ELM model with 800 hidden layer nodes achieves the best performance on Function 5. In other cases, 200 hidden layer nodes are enough to obtain the best performance for ELM models. One explanation is that, because there are a lot of noises in the datasets, arbitrarily setting the number of hidden layer nodes or simply using a lot of hidden layer nodes in ELM models may cause the models to be under-fitting or over-fitting. Due to the number of hidden layer nodes has a significant impact on the performance of ELM, in practical applications, one recommended is to automatically deduce the initial number of hidden nodes by using incremental constructive methods such as incremental ELM (I-ELM) [10], enhanced incremental ELM (EI-ELM) [20], bidirectional ELM (B-ELM) [21], pruned ELM (P-ELM) [22], error minimized ELM (EM-ELM) [23], etc.. And then optimize the initial model to obtain the optimal network model.

4.2. Can the empirical randomization range of hidden weights (i.e., [-1, 1]) lead to the optimal performance of ELM?

In many ELM applications, we found that the authors always used the empirical randomization range [-1, 1] as the default selection range of hidden weights without giving any explanation [20, 24, 25]. To test the rationality of this range (i.e., whether this range can always lead to the optimal performance of ELM), we conducted a thorough experiment on the eight benchmark functions described in Table 1. Here, we set the number of hidden layer nodes to 200, set the randomization range of the threshold of hidden nodes to [0, 1], and use *Sigmoid function* as the activation function of ELM. The variables of testing (i.e., the randomization range of the weights between the input layer and hidden layer) are set to [-1, 1], [-10, 10], [-20, 20], and [-30, 30], respectively. The experimental results are shown in Table 3.

Table 3. The performance of ELM based on different randomization ranges of hidden weights.

Dataset	Weights range	Testing time(s)	Standard deviation	Mean	
				Training RMSE	Testing RMSE
Function3	[-1,1]	1.72E-02	1.67E-02	2.76E-01	**4.16E-01**
	[-10,10]	2.28E-02	1.45E-02	2.84E-01	4.20E-01
	[-20,20]	1.78E-02	1.67E-02	2.84E-01	**4.16E-01**
	[-30,30]	1.25E-02	**1.25E-02**	2.83E-01	4.21E-01
Function4	[-1,1]	1.75E-02	**2.92E-06**	1.93E-06	**1.93E-06**
	[-10,10]	1.81E-02	5.41E-06	6.09E-06	6.09E-06
	[-20,20]	2.09E-02	5.65E-06	4.72E-06	4.72E-06
	[-30,30]	1.38E-02	1.25E-02	4.15E-06	4.15E-06
Function5	[-1,1]	1.97E-02	2.79E-04	2.20E-04	3.07E-04
	[-10,10]	1.94E-02	**6.01E-06**	3.80E-06	**3.80E-06**
	[-20,20]	2.00E-02	1.11E-03	2.43E-06	5.73E-04
	[-30,30]	1.75E-02	1.23E-03	2.17E-06	7.86E-04
Function6	[-1,1]	1.69E-02	5.77E-06	6.75E-06	5.77E-06
	[-10,10]	1.28E-02	8.48E-06	6.35E-06	6.35E-06
	[-20,20]	1.81E-02	**5.00E-06**	4.36E-06	4.36E-06
	[-30,30]	1.22E-02	5.27E-06	4.27E-06	**4.27E-06**

As shown in Table 3, the randomization range of the weights between the input layer and hidden layer has significant impact on the performance of ELM. In addition, we found that the empirical randomization range (i.e., [-1, 1]) may not lead to the optimal performance of ELM models, e.g. Function 3/5/6 cases, which implies that many existing ELM models are likely to achieve better performance by tuning the randomization range of hidden weights.

4.3. Can the empirical randomization range of the hidden threshold (i.e., [0, 1]) lead to the optimal performance of ELM?

Similar to Section 4.2, here we test that whether the most common randomization range (i.e., [0, 1]) of the threshold of hidden nodes can to lead to the optimal performance of ELM. In this experiment, the number of hidden layer nodes is set to 200, the randomization range of the hidden weights is set to [-1, 1], and *Sigmoid function* is used as the activation function of ELM. The variables of testing (i.e., the randomization range of the threshold of hidden nodes) are set to [0, 1], [0, 5], [0, 10], and [0, 15], respectively. The experimental results are shown in Table 4.

Table 4. The performance of ELM based on different randomization ranges of bias.

Dataset	Bias range	Testing time(s)	Standard deviation	Mean	
				Training RMSE	Testing RMSE
Function1	[0,1]	1.56E-02	**4.98E-08**	2.79E-08	3.62E-08
	[0,5]	1.50E-02	9.58E-08	3.11E-08	**3.15E-08**
	[0,10]	1.53E-02	7.61E-07	5.60E-07	5.60E-07
	[0,15]	1.69E-02	6.29E-06	5.98E-06	5.98E-06
Function2	[0,1]	1.59E-02	**6.80E-09**	6.18E-09	**7.33E-09**
	[0,5]	1.56E-02	8.22E-08	5.12E-08	5.12E-08
	[0,10]	1.47E-02	2.34E-06	1.12E-06	1.12E-06
	[0,15]	1.06E-02	4.68E-06	4.86E-06	4.86E-06
Function3	[0,1]	1.72E-02	1.67E-02	2.76E-01	**4.16E-01**
	[0,5]	1.75E-02	**1.47E-02**	2.88E-01	4.31E-01
	[0,10]	1.88E-02	2.74E-02	2.94E-01	4.77E-01
	[0,15]	1.34E-02	8.35E-01	2.94E-01	1.21E+00
Function4	[0,1]	1.75E-02	**2.92E-06**	1.93E-06	**1.93E-06**
	[0,5]	1.84E-02	5.30E-06	4.96E-06	4.96E-06
	[0,10]	1.34E-02	7.41E-06	8.06E-06	8.06E-06
	[0,15]	1.16E-02	1.53E-05	1.63E-05	1.63E-05
Function5	[0,1]	1.97E-02	2.79E-04	2.20E-04	3.07E-04
	[0,5]	2.09E-02	6.16E-05	2.31E-05	3.77E-05
	[0,10]	1.41E-02	6.68E-07	3.76E-07	5.28E-07
	[0,15]	1.53E-02	**2.05E-08**	1.20E-08	**1.34E-08**
Function6	[0,1]	1.69E-02	**5.77E-06**	6.75E-06	**5.77E-06**
	[0,5]	1.09E-02	9.27E-06	1.06E-05	1.06E-05
	[0,10]	1.38E-02	1.40E-05	1.48E-05	1.48E-05
	[0,15]	1.44E-02	1.31E-05	1.42E-05	1.42E-05
Function7	[0,1]	1.69E-02	**6.05E-08**	2.54E-08	**2.60E-08**
	[0,5]	1.44E-02	6.81E-07	1.87E-07	1.87E-07
	[0,10]	2.06E-02	4.23E-06	2.39E-06	2.39E-06
	[0,15]	1.25E-02	7.83E-06	7.19E-06	7.19E-06
Function8	[0,1]	1.50E-02	**1.61E-03**	7.86E-03	1.93E-02
	[0,5]	1.72E-02	2.46E-03	3.65E-03	**1.64E-02**
	[0,10]	1.00E-02	1.85E-02	9.89E-04	2.52E-02
	[0,15]	1.91E-02	1.08E-01	5.64E-04	7.12E-02

From Table 4, we can see that the empirical randomization range (i.e., [0, 1]) may not always lead to the best performance of ELM models, e.g. Function 1/3/5/8 cases. Similarly, it implies that some existing ELM models may achieve better performance by tuning the randomization range of the threshold of hidden nodes.

4.4. The effect of the different activation functions on the performance of ELM.

To test the impact of different activation functions on the performance of ELM model, we conducted a set of experiments. Here, we set the number of hidden layer nodes to 200, set the randomization range of the hidden weights to [-1, 1], and set the randomization range of the threshold of hidden nodes to [0, 1]. The variables of testing (i.e., the functions that used as activation functions) are set to *sigmoid* (*Sigmoid function*), *sine* (*Sine function*), *hardlim* (*Hardlim function*), *tribas* (*Triangular basis function*), and *radbas* (*Radial basis function*), respectively. The experimental results are shown in Table 5.

Table 5. The performance of ELM based on different activation functions.

Dataset	Activation functions	Testing time(s)	Standard deviation	Mean	
				Training RMSE	Testing RMSE
Function1	*Sigmoid*	1.56E-02	**4.98E-08**	2.79E-08	**3.62E-08**
	sine	3.30E-03	1.10E-04	2.61E-03	3.30E-03
	hardlim	9.69E-03	2.57E-04	4.31E-15	2.44E-04
	tribas	1.19E-02	6.40E-03	4.53E-02	4.82E-02
	radbas	8.75E-03	2.48E-03	1.43E-02	1.73E-02
Function2	*Sigmoid*	1.59E-02	**6.80E-09**	6.18E-09	**7.33E-09**
	sine	2.03E-02	1.55E-04	1.30E-03	1.66E-03
	hardlim	1.06E-02	1.55E-04	4.82E-15	7.99E-05
	tribas	1.22E-02	5.59E-03	3.28E-02	3.46E-02
	radbas	1.34E-02	1.38E-03	7.60E-03	9.08E-03
Function3	*Sigmoid*	1.72E-02	1.67E-02	2.76E-01	**4.16E-01**
	sine	2.19E-02	1.63E-02	5.04E-01	6.48E-01
	hardlim	8.44E-03	**1.41E-02**	2.89E-01	4.18E-01
	tribas	8.13E-03	2.45E-02	5.29E-01	7.05E-01
	radbas	1.53E-02	2.14E-02	4.70E-01	6.36E-01
Function4	*Sigmoid*	1.75E-02	2.92E-06	1.93E-06	**1.93E-06**
	sine	1.25E-02	**1.02E-07**	2.04E-06	2.57E-06
	hardlim	1.75E-02	4.30E-05	8.81E-15	6.14E-06
	tribas	1.50E-02	3.07E-04	1.61E-03	1.65E-03
	radbas	1.53E-02	1.04E-05	2.41E-05	2.76E-05
Function5	*Sigmoid*	1.97E-02	**2.79E-04**	2.20E-04	**3.07E-04**
	sine	2.06E-02	5.82E-03	1.41E-01	1.78E-01
	hardlim	1.13E-02	1.06E-03	2.85E-15	2.53E-03
	tribas	1.09E-02	3.08E-01	1.33E-01	2.41E-01
	radbas	1.38E-02	1.31E+02	8.64E-02	3.63E+01
Function6	*Sigmoid*	1.69E-02	5.77E-06	6.75E-06	5.77E-06
	sine	1.38E-02	**5.01E-09**	8.61E-08	**1.12E-07**
	hardlim	6.25E-03	4.43E-05	8.54E-15	6.32E-06
	tribas	1.13E-02	4.56E-05	2.93E-04	3.02E-04
	radbas	1.78E-02	5.55E-07	9.26E-07	1.05E-06
Function7	*Sigmoid*	1.69E-02	**6.05E-08**	2.54E-08	**2.60E-08**
	sine	2.22E-02	2.46E-05	5.46E-04	7.00E-04
	hardlim	1.19E-02	1.78E-04	6.34E-15	1.18E-04
	tribas	1.19E-02	2.25E-02	2.04E-02	2.53E-02
	radbas	1.19E-02	6.44E-04	3.59E-03	4.19E-03

As shown in Table 5, we found that the choice of the activation function has a significant impact on the performance of ELM. In our experiments, the ELM models with *sigmoid* as activation function can achieve the best performance in most cases, while the models with *tribas* as activation function always obtain suboptimal results. One explanation is that the datasets generated by different benchmark functions have different data distributions, and the effect of the feature mappings generated by each data distribution under different activation functions is quite different, which directly affect the performance of ELM. In other words, the *sigmoid* function may perform better nonlinear transformations and obtain more meaningful feature mappings than the *tribas* function on these benchmark functions.

5. Conclusions and future work

In this paper, we did a comprehensive study on the relationship between the parameters and the performance of ELM based on eight benchmark functions. Based on the experimental results, we have the following interesting findings.

(1) More hidden layer nodes may not always lead to the better performance of ELM. Arbitrarily setting the number of hidden nodes may cause the ELM model to be under-fitting or over-fitting. One recommendation is to use the incremental constructive method to obtain an initial network model, and then prune the unimportant hidden nodes to obtain the optimal network model.

(2) The empirical randomization range (i.e., [-1, 1]) of the weights between the input layer and hidden layer and the empirical randomization range (i.e., [0, 1]) of the threshold of hidden nodes cannot guarantee better performance for ELM models. One recommendation is to use the grid search or random search to obtain the best parameters for ELM model.

(3) The effects of five commonly used activation functions on the performance of ELM are shown in Table 5. It turns out that the *sigmoid function* always leads to better performance than the *tribas function* on these regression problems.

In the future, we will give a complete theoretical proof for these phenomena and establish the quantitative relationship between the parameters and the performance of ELM. In addition, ELMs with kernel functions and the multi-layer ELMs will be studied. We hope this study could provide guidance for researchers to select appropriate parameters for ELM.

Acknowledgments

This research is supported by the National Natural Science Foundation of China under Grant nos. 61672358.

References

[1] Chauvin Y and Rumelhart DE 1995 *Backpropagation: theory, architectures, and applications* (Hillsdale, NJ: Lawrence Erlbaum)

[2] Schmidt WF, Kraaijveld MA and Duin RPW 1992 Feedforward neural networks with random weights, *Proceedings of 11th IAPR International Conference on Pattern Recognition, Conference B: Pattern Recognition Methodology and Systems*, IEEE, Vol. **2**, pp. 1–4.

[3] Pao YH and Takefji Y 1992 Functional-link net computing: theory, system architecture, and functionalities, *Computer*, Vol. **25**, No. 5, pp. 76–79.

[4] Tang J, Deng C and Huang GB 2016 Extreme learning machine for multilayer perceptron, *IEEE Trans. Neural Netw. Learn Syst.*, Vol. **27**, No. 4, pp. 809–821.

[5] Hu JY, Zhang JS, Zhang CX and Wang J 2016 A new deep neural network based on a stack of single-hidden-layer feedforward neural networks with randomly fixed hidden neurons, *Neurocomputing*, Vol. **171**, pp. 63-72.

[6] Gu Y, Chen YQ, Liu JF and Jiang XL 2015 Semi-supervised deep extreme learning machine for Wi-Fi based localization, *Neurocomputing*, Vol. **166**, pp. 282-293.

[7] Zhang L and Suganthan PN 2016 A comprehensive evaluation of random vector functional link networks, *Inf. Sci.*, Vol. **367**, pp. 1094-1105.

[8] Li M and Wang DH 2016 Insights into randomized algorithms for neural networks Practical issues and common pitfalls, *Inf. Sci.*, Vol. **382-383**, pp. 170-178.

[9] Huang GB, Zhu QY and Siew CK 2004 Extreme learning machine a new learning scheme of feedforward neural networks, *Proceedings of 2004 IEEE International Joint Conference on Neural Networks*, IEEE, Vol. **2**, pp. 985-990.

[10] Huang GB, Chen L and Siew CK 2006 Universal approximation using incremental constructive feedforward networks with random hidden nodes, *IEEE Trans. Neural Netw.*, Vol. **17**, pp. 879–892.

[11] Huang GB, Zhu QY and Siew CK 2006 Extreme learning machine: theory and applications, *Neurocomputing*, Vol. **70**, pp. 489–501.

[12] Mohammed AA, Minhas R, Wu QMJ and Sid-Ahmed MA 2011 Human face recognition based on multidimensional PCA and extreme learning machine, *Pattern Recog.*, Vol. **44**, No. 10, pp. 2588–2597.

[13] Lu J, Zhao JW and Cao FL 2014 Extended feed forward neural networks with random weights for face recognition, *Neurocomputing*, Vol. **136**, pp. 96-102.

[14] Wong PK, Yang ZX, Vong CM and Zhong JH 2014 Real-time fault diagnosis for gas turbine generator systems using extreme learning machine, *Neurocomputing*, Vol. **128**, pp. 249-257.

[15] Bakri AE, Koumir M and Boumhidi I 2016 Extreme learning machine for fault detection and isolation in wind turbine, *International Conference on Electrical and Information Technologies (ICEIT)*, IEEE.

[16] Guo LL and Ding SF 2015 A hybrid deep learning CNN-ELM model and its application in handwritten numeral recognition, *Journal of Computational Information Systems*, Vol. **11**, No. 7, pp. 2673-2680.

[17] Xie ZG, Xu K, Shan W, Liu LG, Xiong YS and Huang H 2015 Projective Feature Learning for 3D Shapes with Multi-View Depth Images, *Comput. Graph. Forum*, Vol. **34**, No. 7, pp. 1-11.

[18] He YL, Wang XZ and Huang JZX 2016 Fuzzy nonlinear regression analysis using a random weight network, *Inf. Sci.*, Vol. **364-365**, pp. 222-240.

[19] Huang GB 2015 What are extreme learning machines? Filling the gap between frank Rosenblatt's dream and John von Neumann's puzzle, *Cogn. Comput.*, Vol. **7**, No. 3, pp. 263-278.

[20] Huang GB and Chen L 2008 Enhanced random search based incremental extreme learning machine, *Neurocomputing*, Vol. **71**, pp. 3460-3468.

[21] Yang YM, Wang YN and Yuan XF 2012 Bidirectional extreme learning machine for regression problem and its learning effectiveness, *IEEE Trans. Neural Netw. Learn Syst.*, Vol. **23**, No.9, pp.1498-1505.

[22] Rong HJ, Ong YS, Tan AH and Zhu Z 2008 A fast pruned-extreme learning machine for classification problem, *Neurocomputing*, Vol. **72**, pp. 359-366.

[23] Feng GR, Huang GB, Lin QP and Gay R 2009 Error minimized extreme learning machine with growth of hidden nodes and incremental learning, *IEEE Trans. Neural Netw.*, Vol. **20**, No. 8, pp. 1352-1357.

[24] Zhao XM, Cao WP, Zhu HY, Ming Z and Ashfaq RAR 2016 An initial study on the rank of input matrix for extreme learning machine, *Int. J. Mach. Learn. Cybern.*, pp. 1-13.

[25] Salam MA, Zawbaa HM, Emary E, Ghany KKA and Parv B 2016 A hybrid dragonfly algorithm with extreme learning machine for prediction, *International Symposium on INnovations in Intelligent SysTems and Applications (INISTA)*, IEEE.

IOP Conf. Series: Materials Science and Engineering **261** (2017) 012003 doi:10.1088/1757-899X/261/1/012003

Fault-tolerant Control of a Cyber-physical System

Rusu-Both Roxana and Dulf Eva-Henrietta

Automation Department, Technical University of Cluj-Napoca, Romania

*roxana.both@aut.utcluj.ro, eva.dulf@aut.utcluj.ro

Abstract. Cyber-physical systems represent a new emerging field in automatic control. The fault system is a key component, because modern, large scale processes must meet high standards of performance, reliability and safety. Fault propagation in large scale chemical processes can lead to loss of production, energy, raw materials and even environmental hazard. The present paper develops a multi-agent fault-tolerant control architecture using robust fractional order controllers for a (^{13}C) cryogenic separation column cascade. The JADE (Java Agent DEvelopment Framework) platform was used to implement the multi-agent fault tolerant control system while the operational model of the process was implemented in Matlab/SIMULINK environment. MACSimJX (Multiagent Control Using Simulink with Jade Extension) toolbox was used to link the control system and the process model. In order to verify the performance and to prove the feasibility of the proposed control architecture several fault simulation scenarios were performed.

1. Introduction

Cyber-physical systems represent an interconnection between process computer and physical systems through communication, sensor and actuator technologies. The application domains are rapidly increasing like: medical, automotive, aircraft and even military systems. However, in spite all the advantages introduced by this control approach there is also a key issue: the fault control system [1, 2]. Industrial chemical processes, like the (^{13}C) cryogenic isotope separation column cascade are even more difficult to control due to the high complexity and nonlinear dynamics. The research regarding trains of cryogenic isotope separation columns, is scarce, because is a very unique equipment, only few such columns exist in the world. The automation of the isotope separation cascade system is a lot more complicated than in the case of a single separation column making necessary the use of a distributed control system due to many interconnections amongst the columns. Also, distributed control systems use specific real time networks and communication protocols that may introduce variable time delays and even data consistency problems which can lead in extreme cases to the instability of the system making fault control system mandatory. Even if multi-agent systems (MAS) were widely used to study distributed control and collaborative tasks for complex systems there is still the need for fault tolerance in order to increase dependability even in the presence of faults [3, 4]. The research community proposed various algorithms and architectures for fault tolerant solutions. In [5] Khalili et all. proposed an adaptive fault-tolerant control (FTC) scheme for a class of nonlinear uncertain multi-agent systems. Bora and Dikenelli introduce a replication-based approach, a centralized self-adaptive fault tolerant approach that exploits a feedback control loop in [6]. The solution proposed in this work is a fault-tolerant distributed control architecture based on fractional order controllers in order to handle the possible system uncertainties exploiting the increased robustness of the fractional order controllers (FOPID) which have two more degrees of freedom [7, 8]. This approach was tested successfully by the authors in prior studies in which fractional order controllers were implemented for control of a single isotope separation column [9-11]. The only downside of the proposed approach is represented by the increased complexity in FOPID controllers' parameter tuning procedure by using optimization techniques. The fault tolerant architecture concerns mainly the interactions between the subsystems by introducing a supervisor layer that identifies faults and determines the control configuration to keep the closed loop system performances in an acceptable domain. In the present work, fault tolerant multi-agent system FTMAS was developed and implemented in JADE and then it was linked with MATLAB/Simulink, using Multiagent Control Using Simulink with Jade Extension (MACSimJX). The paper is structured into four sections. After the introductory part, section 2 presents the plant characteristics. The third section is focused on the proposed fault tolerant multi-agent system design and implementation. The work ends with concluding remarks presented in section 4.

2. Plant Characteristics

The (^{13}C) isotope has the most uses in a broad range of industrial fields like medicine, biology, chemical engineering and many more: it can be used as a valuable tracer in biological, chemical and environmental science being very useful in medical diagnostics [10]. The main problem is that the natural abundancy of the (^{13}C) isotope is very low (1.1 at.%) which can be raised by cryogenic distillation of the carbon monoxide. This method is based on the higher isotope effect, [12] and on the isotopes nuclear properties being implemented for the available pilot plant developed at NIRDIMT – Cluj-Napoca, Romania. A single column of approximatively 7m height and having a diameter or about ~16mm was able to raise the (^{13}C) concentration only up to (8-10) at.%.

The plant considered in this work consists of three (^{13}C) cryogenic isotope separation columns connected in series, in a cascade, Figure 1a using a common condenser (cooled with liquid nitrogen) with condensation spaces for each of the three columns and a common thermal insulation. Each column is an independent complex nuclear process itself being used for the production of (^{13}C) carbon stable isotope. Each column is packed with Heli-Pack stainless steel wire of 1.8x1.8x0.2 mm in order to increase the contact surface area between the two counteracting flows and is placed inside a vacuum jacket because the operating temperature is around -190°C. The pilot plant is designed in such way that permits the individual operation of the columns as well as operation in series or parallel. However, this technological solution increases the complexity of the control system greatly, the number of variable (monitored or controlled) triples hence a complex control method is needed. To grasp the complexity of the whole process it is worth to mention that start-up procedure of the whole process takes around 3-4 days, and after that the isotope the cascade has to operate for months in order to reach the maximum percentage of (^{13}C). From a fault tolerant control system point of view, the most important objective is to keep the product at the end of the third column as pure as possible while also keeping the whole process in the steady state domain. There are several types of faults that can occur like: plant fault, sensor fault, actuator fault and agent failure. For this particular process in the sensor faults can be included for each column: the temperature sensor in the boiler, the pressure transducers, and the flowmeters. The actuator faults include the flow controllers – important for the proper operation of the cascade, heater present in the boiler of each column, recirculating pumps, vacuum pump for the thermal insulation jacket. In the plant fault category can be included the liquid nitrogen level in the condenser, loss of electrical power and column construction fault. Based on various experimental trials it was concluded that the boiler resistor electrical power supply has a high influence on the isotope separation process. An excessive boiling of the liquid CO may lead to the column flooding compromising in this manner the entire separation process. Hence the present work is focused only on a fraction of the entire control system: the boiler resistor electrical power supply control for all the columns in the cascade.

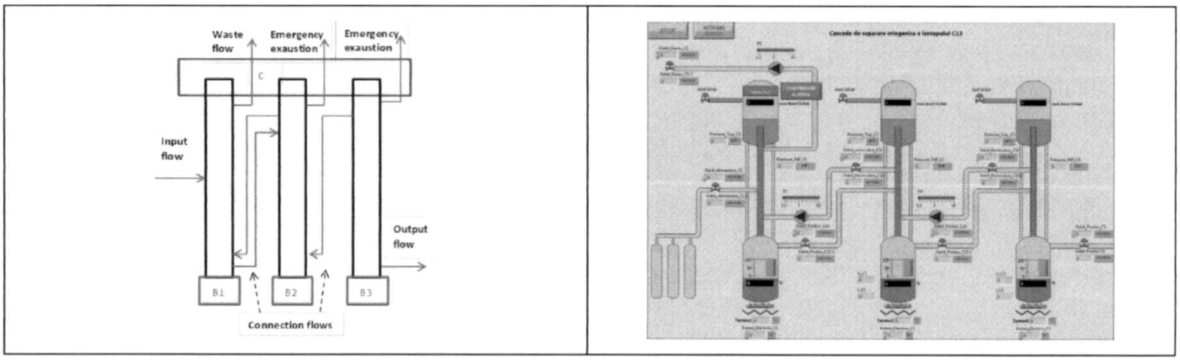

Figure1. a) Schematic diagram of the isotopic separation cascade; b) The current monitoring, control and alarm interface of the isotope separation cascade system

3. Fault-tolerant Multi-Agent Fractional order Control System

Based on the extensive work performed by researchers in the field of fault tolerant control (FTC) it can be stated that FTC is a complex combination of the fault detection and isolation scheme (FDI), robust control and reconfigurable control [13]. The FDI scheme is responsible with fault detection and location in the event of a fault/failure of a sensor, actuator or agent so that corrective action can be made to eliminate or minimize the effect on the overall system performance. In this work a non-model based FDI [13] scheme is implemented based on trend analysis monitoring. In order to detect the faults in time, a fault detection function was developed to help predict the behavior of the process; hence possible faults/failure may be detected and avoided. This function calculates a moving average of a signal by taking a sample every 30 seconds for 15 minutes and then calculates the average. The resulted average is then compared with the previous average and if their difference is greater or equal than the threshold it indicates possible fault. This fault detection function is implemented on the following parameter differential pressure, boiler temperature and on the boiler liquid CO level for each column. Once the fault is detected a visual and audible alarm is activated, Figure 1b and the fault tolerant multi-agent control system can take the necessary corrective actions by modifying the necessary set-point for certain flow control loops and putting some equipment in their safe state in order to keep the process in the steady state regime and preserve the obtained (^{13}C) isotope concentration at the output of the third column.

3.1. The FT-MAS architecture

The chosen platform to implement the proposed fault tolerant multi-agent fractional order control system is the Java Agent Development Environment (JADE) for two reasons: it is an open source platform and it is developed in accordance with the FIPA (Foundation for Intelligent Physical Agents) standards. Because in industrial practice it is desirable to have distributed control the proposed control architecture is a multi-agent based fault-tolerant fractional order control system which consists of the following intelligent agents: one agent for each column (three autonomous agents) and a supervisor agent. Each column agent is responsible for one column, controlled with a fractional order controller. Each column agent has the information corresponding to its assigned column, and has the capability to communicate it to the other agents, receiving theirs in return. The supervisor agent is responsible for the exchange of the data between the column cascade system and the column agents and also with monitoring and the control of the set-point levels for each column agent in order to reach the imposed output (^{13}C) isotope concentration.

JADE is a distributed platform where software agents run and each instance of a runtime environment is called a container. The main container houses three platform key components: the agent management service (AMS), the directory facilitator (DF) and the agent communication channel (ACC) [13]. The AMS maintains a directory with agent identifiers (AIDs) and agent states and the DF facilitates the discovery of network available agents based on their provided services [13]. So it is necessary to effectively manage a possible main container failure to ensure the platform remain fully operational. This objective can be achieved by main container replication and DF persistence [13]. The mathematical model of the (^{13}C) isotope separation column cascade was implemented in Matlab/SIMULINK, but the S-functions available in Matlab/SIMULINK are unable to handle multiple threads of execution, which represent a critical characteristic of MAS. Hence, MACSimJX, was used to act as a middleware between SIMULINK models and the agents. It presents client-server architecture, separating the MAS from SIMULINK [14] being able to overcome the multiple threads of execution problem bringing MAS closer to the physical model. The proposed fault tolerant multi-agent fractional order control architecture is presented in Figure2.

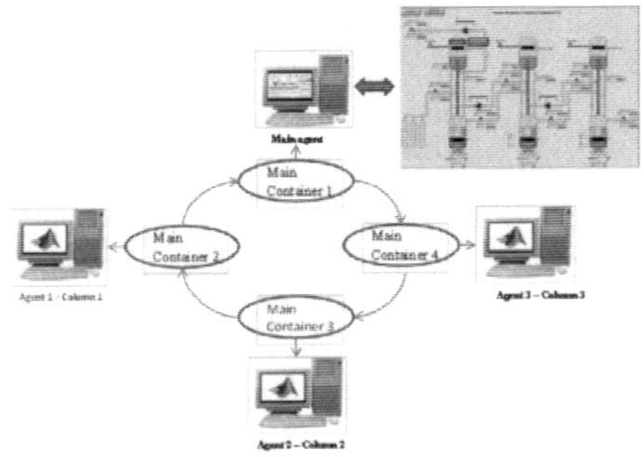

Figure 2. Multi-agent fault tolerant control architecture

3.2. Robust Fractional Order Controller Design

Even if H_∞ control is considered to be one of the most popular robust control methodologies for passive fault tolerant control, the present work tries to improve the robustness of the developed fault tolerant multi agent control system by using a fractional order controller as a behavior method for each column agent. The fractional order controller is practically a generalization of classical PID controller having the following transfer function [7]:

$$H_{FO-PID}(s) = K_P\left(1+\frac{K_i}{s^{\lambda}}+K_d s^{\mu}\right), \quad \lambda, \mu \in \Re \tag{1}$$

where K_P, K_i, K_d are the classical tuning parameters of a PID controller.

The complexity of the controller is increased by the introduction of two extra tuning parameters λ and μ, the integration and differentiation orders. The difficulty is in fact represented by the system of nonlinear equations that need to be solved in order to determine the tuning parameter values. The nonlinear equation system is derived from the explicit form of the imposed performance criteria expressed in frequency domain like: magnitude at gain crossover frequency, phase margin, robustness to plant uncertainties, high frequency noise attenuation and sensitivity functions [8]. Previous works [9-11] were focused on the development of accurate nonlinear mathematical process model which was used together with experimental data in the validation process of an operational model more fitted for further use in control system design. Several fractional order controllers were also developed in previous works but for only one isotope separation column. Hence the controller parameters are determined using the operational process model developed in [11] and based on the chosen imposed performance criteria: the magnitude at the gain crossover frequency, phase margin and open loop phase derivative in order to ensure a certain closed loop settling time, a certain closed loop overshoot and robustness to process gain variations:

$$\left|H_R(j\omega_{gc})\cdot H_P(j\omega_{gc})\right| = 0dB\,, \ \arg\left(H_R(j\omega_{gc})\cdot H_P(j\omega_{gc})\right) = -\pi+\varphi_m\,, \ \left.\frac{d\arg\left(H_R(j\omega)\cdot H_P(j\omega)\right)}{d\omega}\right|_{\omega=\omega_{cg}} = 0 \tag{2}$$

$H_P(s)$ represents the process transfer function and $H_R(s)$ represents the controller transfer function. The performance specifications considered for the present case study are gain crossover frequency of 15 [rad/sec], phase margin of 90° and robustness to gain variations. Usually, optimization techniques are used to determine the tuning parameters of the fractional order controllers. A useful tool to determine the FO controller can be the Particle Swarm Optimization Method. In this particular case, the proposed method is a modified Particle Swarm Optimization (PSO) bio-inspired algorithm based on a technique developed by Eberhart and Kennedy [15] that tries to mimic a group communication process which can be observed when swarms flock, migrate, forage. The main objective is to efficiently search by swarming the best solution in the solution plane. The fitness function is used to evaluate the performance of each particle to determine if the best solution is achieved. Based on the considered imposed performance the fitness function can be expressed as follows, where $x=[K_P \quad K_i \quad K_d \quad \lambda \quad \mu]^T$ are the controller parameters :

$$CF=|f_1(x)|+|f_2(x)|+|f_3(x)| \tag{3}$$

$$f_1=|C_F(j\omega_{gc})\cdot G_P(j\omega_{gc})|,\ f_2=arg(C_F(j\omega_{gc})\cdot G_P(j\omega_{gc}))+\pi-\varphi_m,\ f_3=\left.\frac{d\,arg(C_F(j\omega)\cdot G_P(j\omega))}{d\omega}\right|_{\omega=\omega_{cg}} \tag{4}$$

In order to determine the controller tuning parameter by using the above described particle swarm optimization method an initial population size of 50 particles was considered. Also the inertia weight considered parameters were m=0.9; n = 1.01; p = 1.1; q = 0.051 and r = 1.01. As a result the following controller parameters were determined: K_p=71.51, K_d=0.012, K_i=16.5, λ=0.87 and μ=0.1. The obtained fractional order controllers were implemented using the Oustaloup Recursive Approximation method [9].

3.3. Simulation results and performance analysis

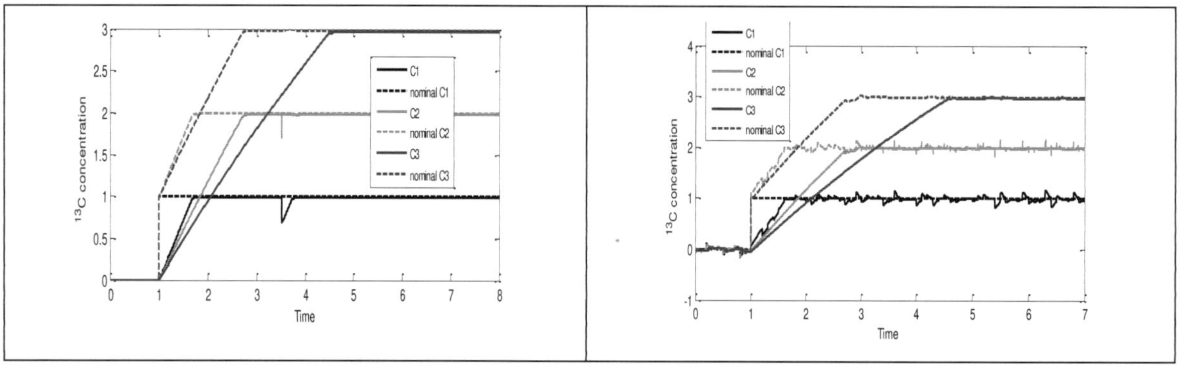

Figure 3. (^{13}C) isotope concentration evolution for all three columns considering: a) a plant fault for the first column; b) sensor measurement error

The next step after the development of any control system is to assess its performance previously to the practical implementation step. Usually, it is important to test the control system capability to counteract the effects of several disturbances and also to test the reference tracking capabilities. An important source of disturbances is represented by various faults, hence, the developed control architecture, Figure2, was used to prove the capabilities of the proposed fault-tolerant multi-agent fractional order control system to handle various faults through simulation. The considered simulation scenarios were inspired by real operation of the (^{13}C) isotope separation cascade. The first considered fault case scenario is a plant fault at t=3.5[sec] on the first column, Figure3a. The disturbance effect is rejected and only the next in line column feels a small part of the effect. The second considered scenario consists in a simple sensor fault for the first column. The results presented in Figure3b show that the performance of the control system decreases but is still in acceptable limits ensuring the stability of the overall process. The presented simulation results clearly demonstrate the capability of the multi-agent fault tolerant control system to counteract and isolate the effects of various faults proving the feasibility of the proposed control system.

4. Conclusions

This work deals with the development and the implementation of a multi-agent system (MAS) to achieve the efficient, fault tolerant control of the (^{13}C) isotope separation column cascade, a highly critical cyber-physical system. The developed controllers are designed to be robust to process gain variations by means of fractional calculus, while the fault tolerance of the system is ensured by communication between the distributed control system nodes. In this way the roles and responsibilities of the faulty element can be overtaken by a neighboring agent. The process mathematical model is implemented in MATLAB/Simulink platform and it is linked with JADE using MACSimJX for realizing the practical implementation of MAS. The simulation results demonstrate the capability of the multi agent system to counteract and isolate the effects of various faults proving the feasibility of the proposed control system. Future works refers to the implementation of such strategy using dedicated software for multi-agent systems based on bio-inspired approach.

Acknowledgment

This work was supported by a grant of the Romanian National Authority for Scientific Research and Innovation, CNCS – UEFISCDI, project number PN-II-RU-TE-2014-4-1465, contract number 38/2015.

References

[1] Duang K, Zhang W 2016 Event-triggered fault-tolerant control for networked systems with dynamic quantiser *IET Control Theory & Applications*, Volume: 10, Issue: 9, 2016
[2] Shahriari-kahkeshi M., Sheikholeslam F. 2016 Active fault tolerant control design for uncertain strict-feedback non-linear systems, *IET Control Theory & Applications,* pp. 1205 – 1214,
[3] Robinson C 2008, Decentralised Data Fusion Using Agents, The University of York, PhD Thesis
[4] Kadera P. 2015 Methods for Development of Industrial Multi-Agent Systems, Czech Technical University in Prague, Doctoral Thesis
[5] Khalili M. et all 2015 Distributed Adaptive Fault-Tolerant Control of Uncertain Multi-Agent Systems, *IFAC-PapersOnline*, vol.48, pp 66-71
[6] Bora S and Dikenelli O. 2016 A centralized self-adaptive fault tolerance approach based on feedback control for multiagent systems, *Turk J Elec Eng & Comp Sci (2016)* 24: 4707 – 4723
[7] Podlubny I 1999 Fractional-order systems and PIλDδ controllers, *IEEE Trans. On Automatic Control*, vol.44, no. 1, pp. 208-213
[8] Monje CA, et.al. 2008 Tuning and auto-tuning of fractional order controllers for industry applications, *Control Eng. Practice*, 16
[9] Both R, Dulf E.H., Muresan C.I. 2014 Fractional order PD controller applied to the Boiler of a ^{13}C Cryogenic Isotope Separation Column, *8th European Nonlinear Dynamics Conference*,
[10] Both R, Dulf E.H., Feştilă C, Agachi P.S. 2012 Robust control of a Catalytic 2 Ethyl-hexenal hzdrogenation reactor, *Journal of Chemical Engineering Science* . 2012;74:300–309.
[11] Both R, et all 2016 Advanced control of a complex chemical process, *Braz. J. Chem. Eng.* vol.33 no.1 São Paulo Jan.Mar.
[12] Axente D, et. al. 1994 Isotope Separation 15N, 18O, 10B, 13C by isotopic exchange, in Romanian, Casa Cartii de Stiinta, 1994
[13] Alwi H et all. 2011 Fault detection and Fault-tolerant control using sliding modes, Springer, ISBN: 978-0-85729-649-8
[14] Robinson C et all. 2010 MACSimJX: A Tool for Enabling Agent Modelling with Simulink Using JADE, *Journal of Physical Agents*

IOP Conf. Series: Materials Science and Engineering **261** (2017) 012003 doi:10.1088/1757-899X/261/1/012003

[15] Li Y., Wei Y., Chu Y. 2015 Research on Solving Systems of Nonlinear Equations Based on Improved PSO, *Mathematical Problems in Engineering*, Vol. 2015, DOI: 10.1155/2015/727218

IOP Conf. Series: Materials Science and Engineering **261** (2017) 012004 doi:10.1088/1757-899X/261/1/012004

Fixed Point Learning Based Intelligent Traffic Control System

Wang Zongyao[1], Sui Cong[2] and Shao Cheng[3*]

[1]Surrey International Institute and School of Tourism and Hotel Management, Dongbei University of Finance and Economics, Dalian, China.

[2]School of Finance and Laboratory of Experimental Economics, Dongbei University of Finance and Economics, Dalian, China

[3]Institute of Advanced Control Technology, Dalian University of Technology, Dalian, China

E-mail: *cshao@dlut.edu.cn

Abstract. Fixed point learning has become an important tool to analyse large scale distributed system such as urban traffic network. This paper presents a fixed point learning based intelligence traffic network control system. The system applies convergence property of fixed point theorem to optimize the traffic flow density. The intelligence traffic control system achieves maximum road resources usage by averaging traffic flow density among the traffic network. The intelligence traffic network control system is built based on decentralized structure and intelligence cooperation. No central control is needed to manage the system. The proposed system is simple, effective and feasible for practical use. The performance of the system is tested via theoretical proof and simulations. The results demonstrate that the system can effectively solve the traffic congestion problem and increase the vehicles average speed. It also proves that the system is flexible, reliable and feasible for practical use.

1. Introduction

With the increasingly rapid urbanization, traffic problem has become a big issue for many countries. Unreasonable road resource usage causes traffic congestion in almost every city around world. The governments make great efforts to develop public traffic system, integrate different types of traffic systems and constraint usage of personal vehicles. But these approaches cannot effectively solve the traffic congestion problem. Firstly, the traffic development is tightly constrained by the fixed urban layout. Secondly, the urban traffic developing speed cannot catch up with the increasing speed of vehicles [1]. So governments utilize intelligence technology to solve urban traffic congestion problems [2]. Intelligence traffic control technology focuses on maximizing the road resources usage rather than "constructing new roads". This paper presents a multi-agent based intelligence traffic control system.

This system utilizes Fixed Point Theorem to adjust traffic flow density. This method can effectively maximize the road resources usage and avoid traffic congestion. As a distributed system, the intelligence traffic control is more flexible and more robust. No central control is needed to manage the system. The system applies a simple algorithm which is easy to build. And it can be easily integrated into existing traditional traffic control system.

2. Literature review

The traffic congestion commonly happens around road intersection. So controlling the traffic light signal at intersection is the key to solve the congestion problem. It has been proved that the traffic flow changing is a pseudo random process [3]. It is very difficult to setup a traffic light control model according to traffic flow. So traditional traffic lights often use fixed time period. Namely, the time periods of green light and red light are fixed. Furthermore, the traditional traffic lights are isolated individuals. The lack of communication and cooperation is also an important factor of traffic congestion. In recent years, researchers have developed many intelligence methods to improve the traffic light control. These intelligence methods are commonly used in small scale traffic network with few intersections [4, 5]. For large scale traffic network, multi-agent technology is commonly used to coordinate the traffic lights inside the network.

Multi-agent technology is a powerful tool to develop large scale distributed systems [6]. It divides a complex network system into a group of intelligence agents. The agents can cooperatively implement a global task which a single agent cannot implement. Multi-agent based traffic control system considers the traffic light controllers as agents. The agents can communicate with each other and cooperatively control the traffic flow according to real time traffic information. Roozemond presents a multi-agent based traffic network control system in paper [7]. The system can autonomously change control rules according to the traffic conditions. The system is composed of traffic light agents, road agents and management agents. The traffic light agents control the traffic flow with assistance of the road agents. The management agents implement global control of traffic light agents. Although the system can achieve cooperation control between agents, it does not clearly state the control rules. The paper [8] introduces an intelligence traffic control system with four kinds of agents: road agents, intersection agents, district agents and central management agents. Through the interaction between agents, the system can implement global optimization control and solve traffic congestion problem. But the centralized control increases the complexity of system and decreases the flexibility and robustness of system.

The main purpose of this paper is to solve traffic congestion caused by unbalanced road resource distribution. The main contributions of this paper are as follows. (1) The presented intelligence traffic control system uses distributed method to control the traffic network. The distributed configuration greatly alleviates the pressure of the large traffic system to the central control. (2) The intelligence traffic control system can autonomously adjust its control rules according to the real-time traffic information (traffic flow density). As applying a simple and practical algorithm (fixed point theorem), the complexity of the system do not increase with the expanding of traffic network. (3) The intelligence traffic control system uses a simple and efficient method to achieve intelligence control. So it is easy to realize and it effectively saves the cost of construction.

3. Traffic Network Control Model

3.1. Traffic Network Model

The model of traffic network is built by using Graph theory [9]. At the intersections, the traffic lights of each direction can be considered as an agent. The agents are connected by roads between them. So topology of a traffic network can be represented by a graph $G = (V; \mathcal{E})$ where V is set of agents and $\mathcal{E}$ is the set of edges. Any two agents can be considered as neighbours if they are connected by a road. If agent j is a neighbour of agent i, we have $j \in N_i$ where $N_i = \{j_1, j_2, \cdots j_n\}$ is neighbour set of agent i.

In order to simplify and do not lose generality, all the roads are bidirectional roads with 6 lanes. When vehicles reach intersection controlled by agent i, vehicles on the right lane will turn to the right, vehicles on the left lane will turn to the left, and vehicles on the middle lane will go straight.

3.2. Traffic Network Model

In actual traffic network, the lengths of roads are different. The number of "accommodated" vehicles is also different from road to road. In such case, the number of vehicles cannot reflect the crowdedness

IOP Conf. Series: Materials Science and Engineering **261** (2017) 012004 doi:10.1088/1757-899X/261/1/012004

of the roads. So a parameter, **traffic flow density**, is defined to evaluate the level of congestion in traffic network. The traffic flow density of a road is defined as: number of vehicles on the road divided by the length of the road. The traffic flow density of road controlled by agent i can be expressed as x_i.

The traffic flow density of agent i is controlled by green light time period of agent i (T_i). Obviously, the longer time period of the green lights can release more vehicles on the road. Accordingly, the traffic flow density will be decreased. The traffic flow density released by agent i can be defined as $y_i = f(T_i)$. The proposed intelligence traffic control system aims to find the optimal released flow density for all agents based on real - time traffic information. If optimal released flow density of agent i is determined, the green light time period of agent i can be determined by $T_i = f^{-1}(y_i)$.

In matter of fact, the traffic congestions often happen at a few intersections in the traffic network. While these busy intersections often become bottlenecks of the entire traffic network. To solve the problem, we can maximize the usage of all the roads in the traffic network. To achieve the purpose, the proposed traffic control system will evenly distribute the vehicles to all the roads. Namely, the traffic flow density of each road will be controlled to reach a same value. Based on this idea, the purpose of optimization is to minimize $min(var(x))$, where $var(x)$ is variance of traffic flow densities of all the roads. Clearly, the best condition of the traffic network system is $var(x) = 0$. Namely, $x_1 = x_2 = \cdots = \bar{x}$ where $\bar{x}$ is the average traffic flow density of all roads in the traffic network.

In a traffic network with n agents, the agent j is one of the neighbours of agent i. y_{ij} is traffic flow density transferred from agent i to agent j. Consider all the neighbour pairs in the traffic network, a transferred traffic flow density matrix can be written as Y. The $(i,j)^{th}$ entry of matrix Y can be expressed as

$$\begin{cases} Y_{ij} > 0 & (j \in N_i) \\ Y_{ij} = 0 & (j \notin N_i \ or \ i = j). \end{cases} \tag{1}$$

So the traffic flow density released by agent i can be written as $y_i = \sum_{j \in N_i} y_{ij}$. And $y = \{y_1, y_2, \cdots, y_n\}$ is traffic flow density vector of the entire traffic network. A weight matrix of traffic flow density is defined as Π. The $(i,j)^{th}$ entry of matrix Π can be expressed as $\Pi_{ij} = y_{ij}/y_i$. Obviously, for any $i \in \mathcal{E}$, $\sum_{j \in N_i} \Pi_{ij} = 1$. According to the definition of traffic flow, we can write the traffic flow of agent i as

$$\sum_{j \in N_i} \Pi_{ji} y_j + x_i - y_i. \tag{2}$$

The intelligence traffic control system aims to average the traffic flow density of all agents. For agent i, the goal of optimization is to keep the traffic flow density at an average value of the whole traffic network:

$$\sum_{j \in N_i} \Pi_{ji} y_j + x_i - y_i = \bar{x}. \tag{3}$$

According to equation (3), the optimal released flow of agent i can be expressed as

$$y_i = max(\sum_{j \in N_i} \Pi_{ji} y_j + x_i - \bar{x}, 0). \tag{4}$$

In an intersection with four traffic lights (controlled by agent 1, agent 2, agent 3, agent 4), only one of the agents is the green light, and the others three are red light. The three agents with red lights have "incoming" traffic flow only. So the traffic flow density of red light agents will increase. The agent with green light has incoming traffic flow and released traffic flow. When the traffic flow density of each agent is equal to or close to the maximum limit (the traffic system is in a crowded state), the released flow of the green light agent will be greater than that of the "incoming" flow. According to equation (4), the traffic flow density of the green light agent i will decrease ($\triangle y_i^1 > \sum_{j\in N_i} \Pi_{ji}\triangle y_j^1$).

Therefore, the optimal control strategy of a local intersection can be set as: first, the agent with maximum traffic flow density is set to green light. This rule reduces the maximum traffic flow density of agents. Traffic flow density of other agents will increase. Second, the released traffic flow density should be limited. The purpose of the local intersection control strategy is to reduce the traffic flow density variance among the agents. Third, the released traffic flow density should be equal to or close to the maximum limit. Namely, the algorithm will not take effect unless there are congestions in the traffic network. This strategy avoids the problems of the central control system. Furthermore, the algorithm is simple and easy to implement. It can be easily built based on traditional traffic network control system.

4. Simulation

In this paper, MATLAB is used to construct the simulation platform. The traffic map around the Forbidden City in Beijing is modelled as the experimental traffic network. The network contains 41 intersections (41*4 agents) and 63 roads. The longest north-south road is 3.7km and the longest east-west road is 6km. Each road is a bidirectional 6-lane road.

4.1. Equilibrium Traffic Network

In the first experiment, we use the traditional traffic control system to control the vehicles. Green time period for each traffic light is set as 25 seconds. Initially, there is no vehicle in the traffic network. From t = 1 seconds, the vehicles randomly enter the traffic network. We define the number of vehicles entering the network per second as CI. In the first experiment, we set CI = 30cars/ seconds. This type of traffic network is also called equilibrium traffic network. Initially, only a small number of vehicles are randomly distributed in a few roads. As time goes by, more and more vehicles rush into the traffic network. Figure 1(a) depicts the traffic network state at t = 800 seconds. Due to the fixed time period of control signal, traditional traffic control system cannot effectively ease the traffic congestion. The vehicles are completely blocked at each intersection. The traffic network presents a paralyzed state. Furthermore, the traffic flow densities of roads are unbalanced. Some roads are idle, while others are congested.

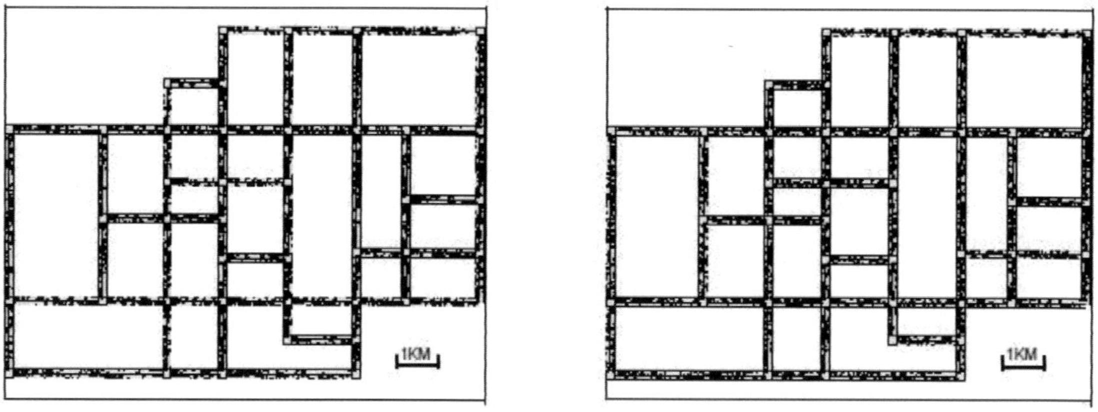

(a) Traditional traffic network at t = 800s (b) Intelligence traffic network at t = 800s

Figure 1. Comparison of Traditional traffic network and intelligence traffic network

Meanwhile, we test the performance of intelligence traffic network control system. At the initial state, the traffic network condition is similar to that of the tradition traffic network. With the passage of time, no congestion happens in the traffic network. Figure 1(b) describes the traffic network in the state of t = 800 seconds. It can be seen that the vehicles are constantly "flooding" the whole traffic network. Although the traffic flow density is very high, there is no obvious congestion at any intersection. The traffic situation is significantly better than the traditional traffic network in figure 1(a).

4.2. Disequilibrium Traffic Network

In the second part of experiment, central area and the eastern area of the traffic network have higher CI (CI = 150cars/s). While the other roads have a lower CI (CI = 30cars/s). This type of traffic network is called disequilibrium traffic network. The disequilibrium traffic network is designed to simulate the real traffic congestion situation. In the real traffic network, traffic congestion commonly happens in certain special area (such as urban centres, downtown, etc). In the experiment of disequilibrium traffic network, the central area and the eastern area are set as crowded areas. We use intelligence traffic control system to control the disequilibrium traffic network. Figure 2 describes the traffic network state at t = 1000s. The figure 2 shows that the centres and eastern part of the traffic network is crowded. But under the control of the intelligence traffic network system, the vehicles can still move in an orderly way. In the surrounding part of the crowded areas, intelligence traffic control system effectively controls the vehicles to bypass the congestion area.

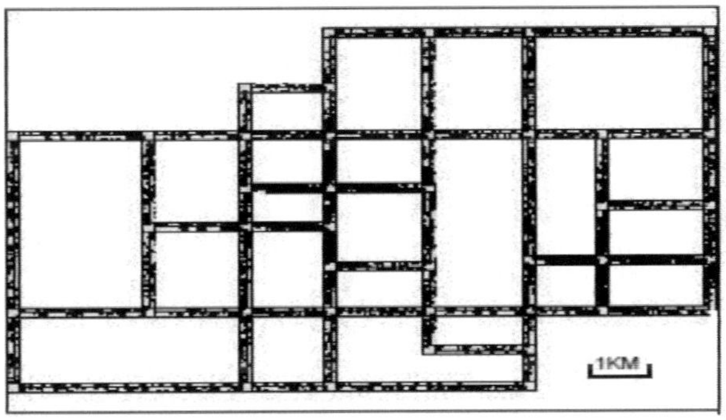

Figure 2. Traffic network status at t = 1000s

To prove the advantages of the intelligence traffic network, we use the traditional control method to control the same type of traffic network and compare the performances of the two control methods. Figure 3(a) and figure 3(b) illustrate the mean traffic flow density and average vehicle speed respectively. The dotted lines represent the traditional control method, and the solid lines represent the parameters of intelligence traffic network control system. The figures prove that the mean traffic flow density of intelligence traffic network is much smaller and more stable than that of traditional traffic network system. Under the control of intelligence traffic network, the vehicle average speed is faster and more stable as well.

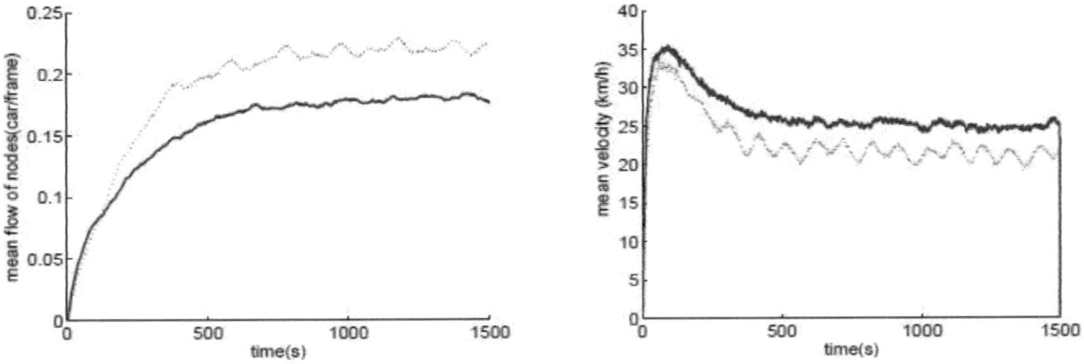

(a) Traffic flow density comparison (b) Average vehicle speed comparison

Figure 3. Traffic network parameters

5. Conclusion

In this paper, an intelligence traffic network control system is established by using multi-agent technology. The algorithm is designed based on fixed point theorem. Through the convergence property of fixed point theorem, the traffic flow density of each road is optimized. The fixed point theorem proves that the intelligence traffic control system can alleviate the traffic congestion and improve the usage of roads. This algorithm adopts the fully distributed structure, which improves the flexibility and robustness of the system. Furthermore, the intelligence system is simple, efficient and easy to implement. It greatly reduces the cost of construction and the complexity of the system.

In the experiment, the graph theory is used to construct the simulation platform of traffic system. In the experiment, we make a comparison between the traditional traffic control system and the proposed intelligence traffic control system. By comparing the parameters, we get conclusion that the intelligence traffic control system can effectively alleviate the traffic congestion and improve the average speed of the vehicles. It also proves that the feasibility of intelligence traffic control system in both equilibrium traffic network and disequilibrium traffic Network.

Acknowledgments

The authors would like to thank the support from National Nature Science Foundation of China (No.61304180, No.71571034), Liaoning province high school talent support program (No.WJQ2015012).

References

[1] M Abdoosa N Mozayania and A L C Bazzanb 2013 Holonic multi-agent system for traffic signals control (Engineering applications of artificial intelligence) Volume 26 pp 1575-1587.
[2] Y Bi D Srinivasan X Lu Z Sun and W Zeng 2014 Type-2 fuzzy multi-intersection traffic signal control with differential evolution optimization (Expert systems with applications) Volume 41 pp 7338-7349.
[3] P G Balaji and D Srinivasan 2011 Type-2 fuzzy logic based urban traffic management (Engineering applications of artificial intelligence) Volume 24 pp 12-22.
[4] P Gregoire C Desjardins J Laumonier and B Chaib-draa 2007 Urban traffic control based on learning agents (IEEE Intelligent transportation systems conference) pp 916-921.
[5] MWeiring 2000 Multi-agent reinforcement learning for traffic light control (Proceedings of the 7th international conference on machine Learning) pp 1151-1158.
[6] B Chen and H Cheng 2010 A review of the applications of agent technology in traffic and transportation systems (IEEE Trans. on intelligence transport systems) Volume 2 pp 485-497.
[7] D A Roozemond 2001 Using intelligent agents for pro-active, real-time urban intersection control (European journal of operational research) Volume 2 pp 293-301.
[8] C Cai and Z Yang 2007 Study on urban traffic management based on multi-agent system (IEEE proceedings of the 6th international conference on machine learning and cybernetics) pp 25-29.
[9] G Chris and R Gordon 2001 Algebraic Graph Theory (Springer-Verlag publisher).

IOP Conf. Series: Materials Science and Engineering **261** (2017) 012005 doi:10.1088/1757-899X/261/1/012005

Deep Learning for Real-Time Capable Object Detection and Localization on Mobile Platforms

F. Particke[1, *], R. Kolbenschlag[1], M. Hiller[1], L. Patiño-Studencki[1] and J. Thielecke[1]

[1]Friedrich-Alexander-Universität Erlangen-Nürnberg (FAU), Information Technologies, Erlangen, Gersmany

[*]florian.particke@fau.de

Abstract. Industry 4.0 is one of the most formative terms in current times. Subject of research are particularly smart and autonomous mobile platforms, which enormously lighten the workload and optimize production processes. In order to interact with humans, the platforms need an in-depth knowledge of the environment. Hence, it is required to detect a variety of static and non-static objects. Goal of this paper is to propose an accurate and real-time capable object detection and localization approach for the use on mobile platforms. A method is introduced to use the powerful detection capabilities of a neural network for the localization of objects. Therefore, detection information of a neural network is combined with depth information from a RGB-D camera, which is mounted on a mobile platform. As detection network, YOLO Version 2 (YOLOv2) is used on a mobile robot. In order to find the detected object in the depth image, the bounding boxes, predicted by YOLOv2, are mapped to the corresponding regions in the depth image. This provides a powerful and extremely fast approach for establishing a real-time-capable Object Locator. In the evaluation part, the localization approach turns out to be very accurate. Nevertheless, it is dependent on the detected object itself and some additional parameters, which are analysed in this paper.

1. Introduction

In future smart factories, every machine will communicate and fuse their information from the physical and virtual world, in order to accomplish fully autonomously their tasks with less effort. Especially smart mobile platforms like robots will enormously lighten the workload and will optimize production processes in Industry 4.0. In order to interact with the environment, the robot has to recognize and localize objects. In addition, especially in traffic control, short response times are important. The contribution of this paper is to establish an object detector and locator for a mobile platform, which is very fast and reliable and works even in unknown environments. With deep learning, currently being state of the art in computer vision, the problem is tackled in this paper by a CNN. Since 2012, every winning team in the ImageNet Large Scale Visual Recognition Challenge [1] has used a convolutional neural net [2], [3]. The recently released YOLO Version 2 (YOLOv2) detection network [4], [5] outperforms current state of the art networks in performance (e.g. VGG Net [6]) and precision (e.g. ZF Net [7]). Additionally, it is empowered to detect almost 9000 different object classes. Hence, it poses as an optimal candidate. As detection networks only locate objects within a camera image, an efficient and real-time capable approach is introduced to translate the powerful detection capabilities of a neural network into the 3D space. It combines detection information of a neural network with depth information from a RGB-D camera mounted on the platform. In order to find the detected object in the depth image, it maps the bounding boxes, predicted by the detection network, to the corresponding regions in the depth image. It provides a powerful and extremely fast approach for establishing a real-time-capable Object Locator.

The paper is organized as follows. In Section 2 the object detection with YOLOv2 is shortly described, in Section 3 the spatial location approach using detection and depth information is proposed. This approach is evaluated in Section 4. Further research is pointed out in Section 5.

2. Object detection with YOLOv2 on the mobile platform

In this Section, the object detection with YOLOv2 (Subsection 2.1) and the integration into the mobile platform (Subsection 2.2) is described. Goal of this section is to summarize the key features in comparison to other convolutional neural networks. For further information the interested reader is referred to [4].

2.1 Object detection with YOLOv2

You Only Look Once (YOLO) is both name and motto of a fast and precise object detection network, which was released in 2016. The system was made to look only once at an image to detect objects. In Figure 1, the basic approach of YOLOv2 is depicted.

YOLOv2 belongs to the class of so called single-shot networks, e.g. SSD [8]. It models object detection "as a single regression problem straight from image pixels to bounding box coordinates" [9], called detection as regression. Thus, it is significantly faster than common networks and has a less complex network pipeline. Based on the entire image, it predicts category scores and box coordinates for a fixed set of bounding boxes, unifying separate components of object detection. YOLOv2 divides an image in a $S \times S$ grid (cf. Figure 1) and predicts B bounding boxes for each grid cell with four coordinates and a confidence score for those boxes. Additionally, each cell predicts C class probabilities. All in all, this leads to a predicted tensor, which is used for regression.

$$SxSx(B \cdot 5 + C) \tag{1}$$

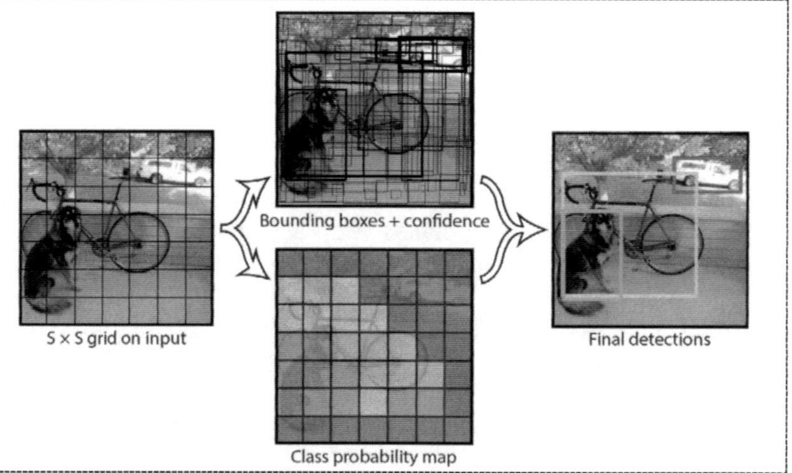

Figure 1. YOLOv2 divides an image into a grid, with each cell predicting a fixed amount of bounding boxes and a class score. Figure is adapted from [10].

Many detection frameworks rely on the powerful and accurate VGG-16 CNN [10] for feature extraction. But it has a major drawback, namely a complex pipeline. This makes it difficult to use in real-time systems, as it requires 30.69 billion floating point operations for a single image in its convolutional layers [4, p.5]. To be real-time capable, YOLOv2 uses an own convolutional neural network called "Darknet". It is less complex but much faster than VGG-16, requiring only 5.58 billion operations while having only a slightly worse accuracy of 2% than VGG-16. In the ImageNet Challenge, it achieved a top-5 accuracy of 88%.

To sum up, the overall advantages for the integration on a mobile platform are

- YOLOv2 outperforms current state of the art networks in performance and precision
- YOLOv2 is able to detect over 9000 different object classes
- YOLOv2 is delivered as an enclosed program. Therefore, it can be used instantly without the necessity of being created and trained

2.2 Integration into a mobile environment

YOLOv2 is integrated into a mobile environment to fuse its detection capabilities with other sensors on a mobile platform. A robot laboratory is used, which provides space and hardware for testing and implementation of YOLOv2. The robot SUMMIT XL, developed by Robotnik [11], serves as a mobile platform. In order to communicate with sensors of the robot, YOLOv2 is combined with the Robot Operating System (ROS). ROS is a framework, providing a communication infrastructure between sensors, which enables them to pass messages to each other. These messages contain sensor data or status information [12]. The RGB-D Camera is used as a sensor for the detection and localization approach.

3. Spatial localization approach

The approach of fusing the classification and detection information of YOLOv2 with the Depth-Image is described in this section. A general overview is depicted in Figure 2. The RGB-Image is used as input of the YOLOv2 algorithm, which outputs the classification and the detection areas of the objects.

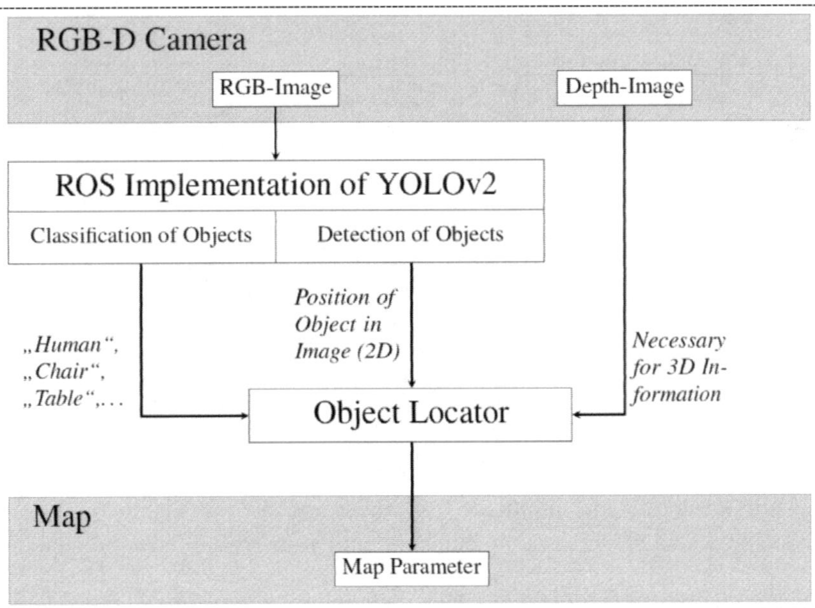

Figure 2. The road map shows the flow of information from the sensor to the created map, which is based on the fusion of deep learning (output of YOLOv2) and the depth-image.

An example of the output of YOLOv2 is presented in Figure 3. Four objects are present in this scenario: chair, cup, sports ball and person. All objects are classified correctly and the bounding boxes are drawn around the objects. The resulting bounding boxes are used for the definition of the regions of interest in the Depth-Image. An example of a Depth-Image with detected bounding boxes is presented in Figure 4.

Figure 3. Output of block "ROS Implementation of YOLOv2" of Figure 2. The detection (bounding boxes) and the classification results are depicted. These results are used for the fusion in the Object Locator.

Figure 4. The depth image, which is enhanced by the detection results of YOLOv2. The bounding boxes select the interesting areas for the Object Locator.

The detected bounding boxes in the Depth-Image are used in the Object Locator. The Object Locator performs the following steps. Firstly, the selected bounding boxes are background filtered. The background is always the surface having the largest distance to the camera. So, all depth values within an interval between the maximum depth and a certain threshold are denoted as "background" and are set to zero. For the following experiments, the threshold was set to 0.05 m. Secondly, the depth values are clustered with K-means. Best results are given for K = 4. Finally, the mean average of the biggest cluster is used for the distance estimate. The classified objects are mapped to an existing database, which represents the current map.

4. Evaluation

The proposed approach of Section 3 is evaluated in dependence of the distance for two different objects, a sports ball and a person. Figure 5 shows the difference between the true depth and the mean predicted depth values from 1m … 5m for the two objects. Due to the fact that the objects are placed in the center of the camera image (x = 0, y = 0), the depth corresponds to the distance between the object and the robot. It can be seen, that in nearer distances the depth values of the sports ball are more accurate than the values of the person. As the surface of the ball is plain, the dominating cluster contains more similar depth values than the surface of a person. As the person is very close to the camera, only the legs of the person are captured by the camera. Therefore, the mean depth value of this cluster slightly differs from the ground truth. The precision of the depth values corresponding to the person gets more accurate at a distance of 2 meters, as there are more parts of the body in the camera view. Hence, the person provides a new kind of surface, which is apparently better located.

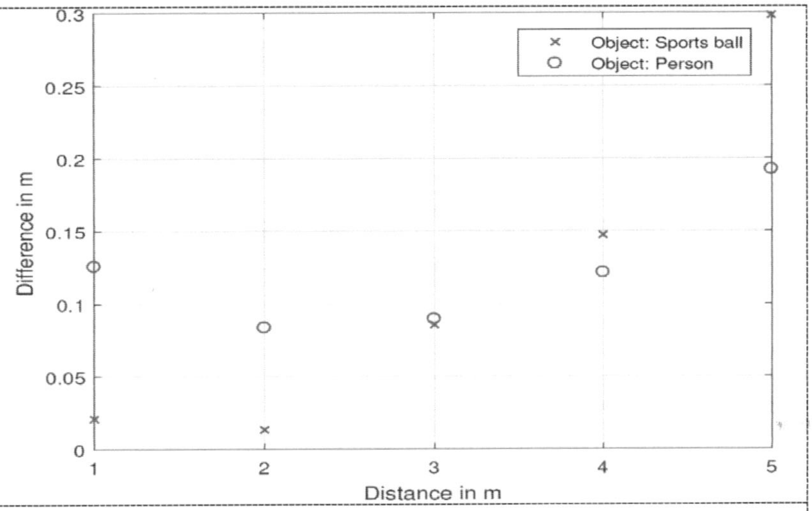

Figure 5. Difference between ground truth and mean predicted depth values in relation to the distance of the object.

With an increasing distance, the depth values of the person are more accurate than the values of the ball, because there are more depth values to evaluate in the chosen cluster and the cluster mean value is less influenced by outliers.

The prediction of the depth values and therefore the localization depends not only on the distance, but also on the surface of the evaluated object. Objects with a plain surface are better localized. In addition, it is shown that the position of near objects is estimated very accurately. Nevertheless, the localization only works well if the correct cluster was chosen. In the case of the ball and person, the number of four clusters was suitable, because no other object was put in the same cluster and every chosen cluster had enough values for a representative centroid. All in all, the approach seems really promising for the fusion of localization with RGB-D and classification with YOLOv2.

5. Conclusion and further work

This paper addressed the object detection within the RGB image of the camera and the fusion of the detection results with depth information. YOLOv2 was used for detection and was integrated into a robotic framework, which made a very fast and powerful object detector available for mobile platforms. It turned out that the detection confidence is very dependent on the number of image points and the object type. Empowered with an internal object detector, new possibilities were given to find the correct depth regions. The difference between the ground truth and the mean predicted depth values in relation to the distance of the object were analysed. It was shown that there is a strong dependence between the distance of the object to the depth sensor and the accuracy of the approach. Nevertheless, the proposed approach showed very promising results. In posterior works, the localization could be further improved by additionally considering spatial information in the clustering process. Furthermore, instead of the K-Means algorithm the X-Means algorithm could improve the proposed Object Locator. It introduces an approach to find the optimal number of clusters in the cluster space by optimizing the Bayesian Information Criterion. Last but not least, it is necessary to extend the Object Locator to provide a statistically proven confidence score.

References

[1] Stanford Vision Lab. Imagenet large scale visual recognition competition (ilsvrc). http://www.image-net.org/challenges/LSVRC/. [Online; accessed 21-June-2017].

[2] Chiyuan Zhang. Mocha.jl: Deep learning for julia. https://devblogs.nvidia.com/parallelforall/author/czhang/, 2015. [Online; accessed 21-June-2017].

[3] Krizhevsky A, Sutskever I, and Hinton G.. Imagenet classification with deep convolutional neural networks. In F. Pereira, C. J. C. Burges, L. Bottou, and K. Q. Weinberger, editors, Advances in Neural Information Processing Systems 25, pages 1097–1105. Curran Associates, Inc, 2012.

[4] Redmon J and Farhadi A. Yolo9000: Better, faster, stronger. arXiv preprint arXiv:1612.08242, 2017.

[5] ThundeRatz robotics. ros_yolo2 - yolov2 integration with ros. https://github.com/ThundeRatz/ros_yolo2, 2017. [Online; accessed 21-June-2017]

[6] Liu, S., & Deng, W. (2015, November). Very deep convolutional neural network based image classification using small training sample size. In *Pattern Recognition (ACPR), 2015 3rd IAPR Asian Conference on* (pp. 730-734). IEEE.

[7] Zeiler, M. D., & Fergus, R. (2014, September). Visualizing and understanding convolutional networks. In *European conference on computer vision* (pp. 818-833). Springer, Cham

[8] Liu W, Anguelov D, Erhan D, Szegedy C, Reed S, Fu C, and Berg A. Ssd: Single shot multibox detector. In European Conference on Computer Vision, pages 21–37. Springer, 2016.

[9] Redmon J, Divvala S, Girshick R, and Farhadi A. You only look once: Unified, real-time object detection. In Proceedings of the IEEE Conference on Computer Vision and Pattern Recognition, pages 779–788, 2016.

[10] Karen Simonyan and Andrew Zisserman. Very deep convolutional networks for large-scale image recognition. arXiv preprint arXiv:1409.1556, 2014.

[11] Robotnik. Mobile robot summit xl | robotnik. http://www.robotnik.eu/mobile-robots/summit-xl/. [Online; accessed 21-June-2016].

[12] Open Source Robotics Foundation. Robot operating system. http://www.ros.org/about-ros/. [Online; accessed 21-June-2017].

IOP Conf. Series: Materials Science and Engineering **261** (2017) 012006 doi:10.1088/1757-899X/261/1/012006

Automated Feature Identification and Classification Using Automated Feature Weighted Self Organizing Map (FWSOM)

Andrew Starkey[1], Aliyu Usman Ahmad[1,*] and Hassan Hamdoun[2]

[1]School of Engineering, University of Aberdeen, Aberdeen AB24 4UE, UK
[2]School of Natural and Computing Sciences, University of Aberdeen, Aberdeen AB24 4UE, UK,

*r01aua14@abdn.ac.uk

Abstract. This paper investigates the application of a novel method for classification called Feature Weighted Self Organizing Map (FWSOM) that analyses the topology information of a converged standard Self Organizing Map (SOM) to automatically guide the selection of important inputs during training for improved classification of data with redundant inputs, examined against two traditional approaches namely neural networks and Support Vector Machines (SVM) for the classification of EEG data as presented in previous work. In particular, the novel method looks to identify the features that are important for classification automatically, and in this way the important features can be used to improve the diagnostic ability of any of the above methods. The paper presents the results and shows how the automated identification of the important features successfully identified the important features in the dataset and how this results in an improvement of the classification results for all methods apart from linear discriminatory methods which cannot separate the underlying nonlinear relationship in the data. The FWSOM in addition to achieving higher classification accuracy has given insights into what features are important in the classification of each class (left and right-hand movements), and these are corroborated by already published work in this area.

1. Introduction

Previous work on the classification of EEG data as described in [1] show that machine learning methods can be used to achieve classification accuracies of up to 97.1%. The work highlighted the difficulty of achieving better classification, and also highlighted that linear classification models performed poorly due to the nonlinear nature of the underlying data. The results from this work [1] show that the performance of the various classification algorithms can vary considerably depending on which features are used for the classification. Whilst for small numbers of features (such as in this work which had 4 features) this is possible and it is not an issue to exhaustively compare each and every combination of features as inputs, for larger datasets this clearly becomes impractical and is therefore otherwise accomplished either by hand or not at all.

In addition, work by [2-4] shows that irrelevant features can result in low classification results. A novel method presented in this paper looks to automatically identify the irrelevant features so that classification performance can be improved. The same EEG data from [1] is used which is publically available data from Physionet [5]. Using this data we are able to compare with this earlier study and highlight that the proposed method is able to improve the classification performance by correctly identifying the relevant inputs and discarding automatically the irrelevant inputs.

2. Methodology

The EEG data will be analysed and classified using a self-organizing map, by SVM linear and cubic methods, and the FWSOM method which are described in the following sections.

2.1. Self-Organizing Map (SOM)
The Self-Organising Map (SOM) is an unsupervised neural network clustering algorithm, referred as Kohonen's SOM [6]. A SOM aims to map data patterns onto n-dimensional grids of neurons; this is inspired by the tendency of the biological neurons that have similar functions stored in the same region of the brain. The SOM's mapping preserves a topological relation by maintaining neighbourhood relations such that patterns that are close in the input space are mapped to units that are close in the output space, and vice-versa.

2.2. Support Vector Machines (SVMs)
SVM is a well-known regression and classification learning algorithm [7, 8]. The basic aim of SVM is finding optimal hyperplanes (which could be linear or nonlinear) that segregate multiple groups.

2.3. Feature Weighted Self Organizing Map (FWSOM)
The FWSOM approach is a novel approach that looks to automatically identify what features are important in a given dataset so that the classification accuracy can be improved.

In the proposed approach, information from what a standard SOM has learnt during training is used to identify what the SOM has seen as important for making decisions and to guide subsequent steps of the training, and to generate individual weightings at a node level which will reduce the importance of inputs that are considered to be irrelevant for that node. It is expected that samples from a given class may spread over multiple nodes (i.e. will not be mapped to a single node) due to any irrelevant features in the dataset. When there are irrelevant inputs in the dataset the SOM will see samples from the same class as different due to these irrelevant inputs.

The analysis of input relevance begins after the SOM is trained. The distances between all nodes in relation to each individual class is calculated in order to identify what is important and irrelevant in the mapping of each class.

For each class, we defined a winner node $N_{winnerC_k}$ as the node with highest number of samples from a given class C_k, neighbouring nodes N_{NeighC_k} as all other nodes with samples from the class mapped, and the distant nodes N_{DistC_k} as other nodes with no samples from the class mapped.

A similarity matrix is calculated as the distance between the mean of the input samples $\mu x_{ck}^{d}$from the class in neighbouring nodes N_{NeighC_k} to corresponding $N_{winnerC_k}$ weight values as (Eqn 1) below;

$$Sim_{Ck}^{d} = \left\| \mu x_{ck}^{d} - w_{N_{winnerC_k}}^{d} \right\| \quad (1)$$

Where d is the input dimension of a given sample x belonging to class ck and w is the weight value of the winner node $N_{winnerC_k}$.

The distance values in the similarity matrix for important inputs for defining the class is expected to be low since it is assumed that these inputs will share similar values for the same class. The distance for the irrelevant inputs is expected to be high since it is assumed that these inputs will have different values in the SOM for the nodes to which the class samples are mapped, and are the reason why the class samples are mapped to different nodes rather than a single node. This process clearly makes the assumption that the class has a single underlying set of features that define the class, and so these assumptions will not be correct when a class is made up of a number of different feature relationships.

In addition, a dissimilarity matrix (Eqn 2) describing the features that can be identified as being different from the current class and all other classes is computed as the distance between the mean of the class samples in the winner node $\mu x_{N_{winnerC_k}}^{d}$ to the weight values of N_{DistC_k} as;

$$Diss_{Ck}^{d} = \left\| w_{N_{DistC_k}}^{d} - \mu x_{N_{winnerC_k}}^{d} \right\| \quad (2)$$

This calculation uses the assumption that the SOM did not map the samples from a given class to these other distant nodes due to the inputs that the SOM sees as different, and therefore the distance values in the dissimilarity matrix for these inputs will be high.

For each class, important inputs are identified as inputs with > 0 percentage change from variance of similarity matrix $SIMC_k$ to the variance of dissimilarity matrix $DISC_k$ as in Eqn 3 below;

$$CkInputs_{roc} = \left(\frac{DISC_k - SIMC_k}{SIMC_k}\right) * 100 \quad (3)$$

3. Experimental Design

3.1. EEG Dataset

EEG signals record the differences of the voltage from two locations on the scalp over time. The EEG signals have an amplitude in the range of 1-100 μv with frequency in the range of 0.5 to 10 Hz [9]. The data recorded and contributed by Physionet using the BCI2000 instrumentation system described in [5] is used.

3.2. Feature Extraction

Frequency bands α and β are chosen for the frequency domain feature extraction because of their ability to distinguish movements in the active state. The delta band is also chosen since research [10] supports that existence of Delta rhythm in the motor cortex within the pre-movement stages (Movement-Related Cortical Potential (MRCP). The EEG data is transformed by extracting Alpha, Beta and Delta band features for 8 electrodes for 6 subjects. For each of the left hand and right-hand tasks separately. The selected electrodes are (C3,C4, Cz, Fc3, Fcz, Fc4, C1, C2). The choice of fewer electrodes is also quite practical and allows the proposed method to be easily incorporated within a portable Wireless EEG system for use in prosthetics applications for example. These electrodes were selected in corroboration with the literature and they include the C3, C4 and Cz electrodes (located on the top of the head) for which distinguishable difference in the Power Spectral Density (PSD) can be observed [11]. Three frequency domain features are extracted from the raw EEG data for each selected electrode by applying Fourier Transform [12]; converting from the time domain to frequency domain characteristics.

3.3. Training and Validation Parameters

The Datasets were separated into training and test sets with the 5-fold cross validation method, accuracy was measured using the confusion matrix [13].

4. Classification Accuracy Results

The table as shown in Table 1 gives the results of classification using the various methods using all inputs, and then also the results using only those inputs identified by the FWSOM process

The FWSOM method returns the identification of irrelevant inputs, and this allows the classification methods to be tested once more with a reduced feature set. These results are also shown in Table 1 for each method, with the relevant features identified by the FWSOM when trained with a 10x10 lattice being used. From an initial input feature set of 24 the FWSOM reduces this to 12 features for alpha band, 73 to 33 for beta band and 20 to 6 for delta band.

The FWSOM has identified the instantaneous power spectrum of the frequency bands at various EEG sampling points as important features for separating the two groups. The location of where the power spectrum sample point features are important is different for the different frequency bands and across the 8 subjects investigated in this paper. The FWSOM has also identified that relative power and peak power features are consistently less important in the classification compared to the instantaneous power spectrum feature. This result is corroborated with those obtained in [11, 14] where it was found that the peak power feature is not a disguisable feature for classification when multiple frequency bands are used and that high and low values of relative power across frequency bands might not be able to be clear distinguisher for EEG data. These results, therefore, show that the FWSOM method has correctly identified the important inputs for this particular dataset.

After application of the FWSOM method, the results of the SVM cubic method have improved, showing that the SVM method is indeed affected by the poor choice of features for training and can struggle to ignore the effect of irrelevant inputs. Interestingly, the SVM linear method shows virtually no change with one result even giving a worse result.

It can be seen that the standard SOM using all inputs gives a poor performance for the small 2x2 lattice size although this does improve when a larger SOM is used of 10x10 lattice. The results also improve when the inputs identified by the FWSOM only are used for training, but 100% diagnosis can still not be achieved.

The results presented in the FWSOM column show that 100% classification performance can be achieved through the use of the feature weighting element that allows the samples to be differentiated from each other based on their class membership more easily as seen in figures 1 to 6.

Table 1. Classification Accuracy Results

EEG Band	**Method**									
	SVM (Linear)		**SVM (Cubic)**		**SOM (2x2 Lattice)**		**SOM (10x10 Lattice)**		**FWSOM (2x2 Lattice)**	**FWSOM (10x10 Lattice)**
	all Inputs	10x10 FWSOM Inputs	all Inputs	10x10 FWSOM Inputs	all Input	10x10 FWSOM Inputs	All Input	10x10 FWSOM Inputs	2x2 FWSOM Inputs	10x10 FWSOM Inputs
α	63%	63%	83%	88%	53%	63%	78%	88%	66%	100%
β	79%	78%	88%	90%	66%	68%	73%	84%	60%	100%
δ	53%	53%	72%	80%	60%	40%	76%	82%	53%	100%

The figures (1-6) show the hits for nodes in the 10x10 lattice from a standard SOM and then for the FWSOM following the process that calculates weights for each node and remaps the samples against these nodes for each of the EEG band datasets. The figures show pie charts for each node that a sample maps to. Therefore, not all nodes are shown in the figures when no samples map to them. The colours of the pie chart relate to the class of the samples that map to it, and are % at a node level.

These graphical representations of the results show clearly that the samples from class 1 (red node) can be represented by a single node, whereas the 2nd class (blue node) require a large number of nodes to allow the variability in this class to be properly represented.

It is clear from the results of the standard SOM shown in the figures that this variability results in the classes being mixed throughout the nodes making classification difficult, and hints at the effect that the irrelevant features are having on the spread of data throughout the SOM. This is likely to be similar to the SVM methods and explains why these methods are also finding classification difficult.

The results of the two different sizes of SOM also indicate that this is an important factor in the analysis process. If the SOM size is too small then the variability in the dataset will not be properly mapped resulting in the ineffective application of the FWSOM method.

Figures (7-12) show bar charts of the equation given in Eqn 3, and illustrate the important indexes for each class found by the FWSOM process (those with a value >0) and the irrelevant indexes for each class (those with a value <0). The analysis for this dataset shows clear differences in the values for the majority of features allowing straightforward identification of both relevant and irrelevant features.

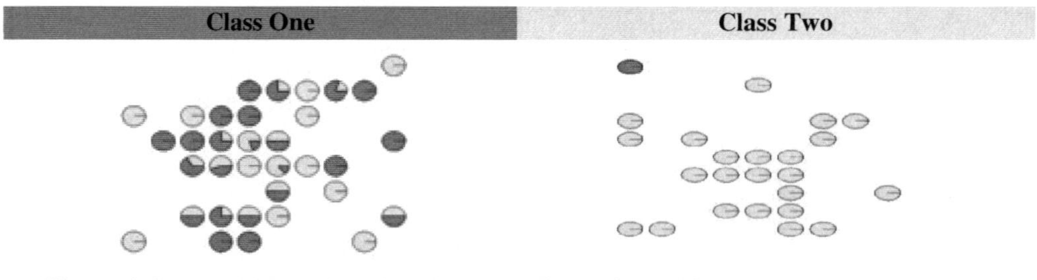

Figure 1. Standard SOM's Samples Hits - alphaBand

Figure 2. FWSOM's Samples Hits - alphaBand

IOP Conf. Series: Materials Science and Engineering **261** (2017) 012006 doi:10.1088/1757-899X/261/1/012006

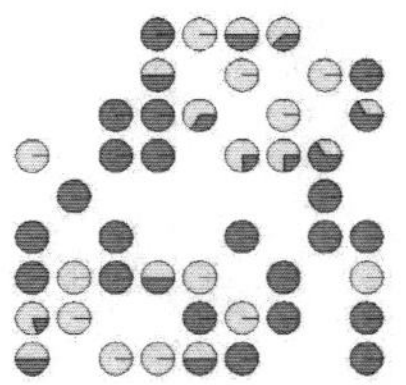

Figure 3. Standard SOM's Samples Hits - betaBand

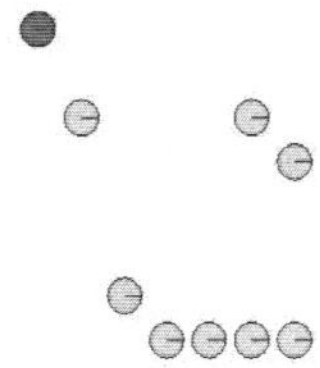

Figure 4. FWSOM's Samples Hits - betaBand

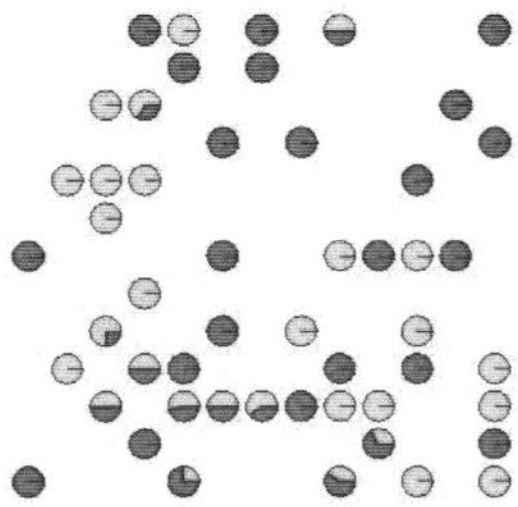

Figure 5. Standard SOM's Samples Hits – deltaBand

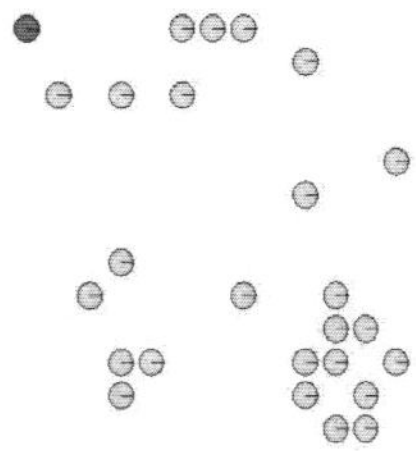

Figure 6. FWSOM's Samples Hits - deltaBand

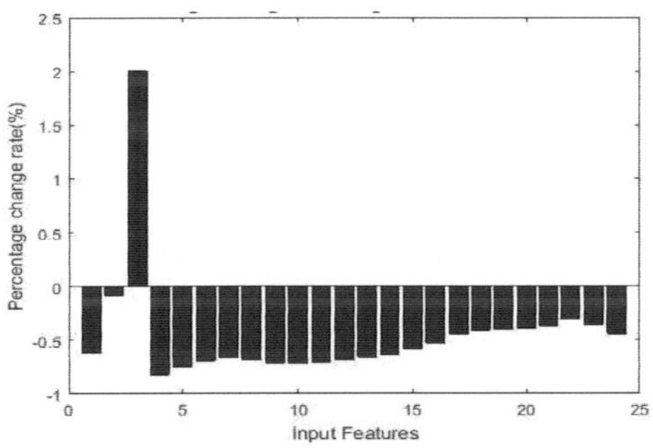

Figure 7: FWSOM's relevant inputs for class 1 - alphaBand

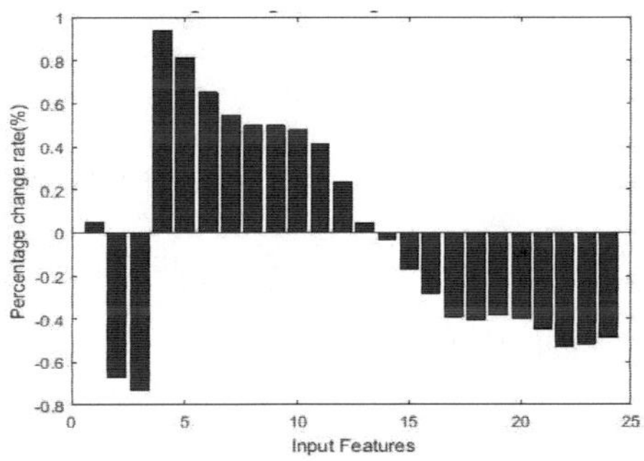

Figure 8: FWSOM's relevant inputs for class 2 - alphaBand

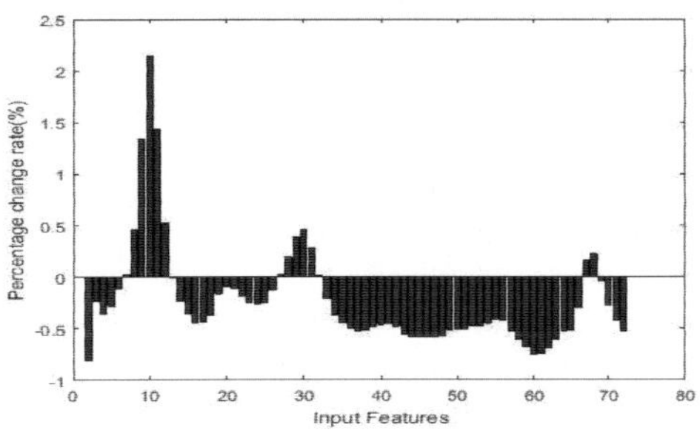

Figure 9: FWSOM's relevant inputs for class 2 - betaBand

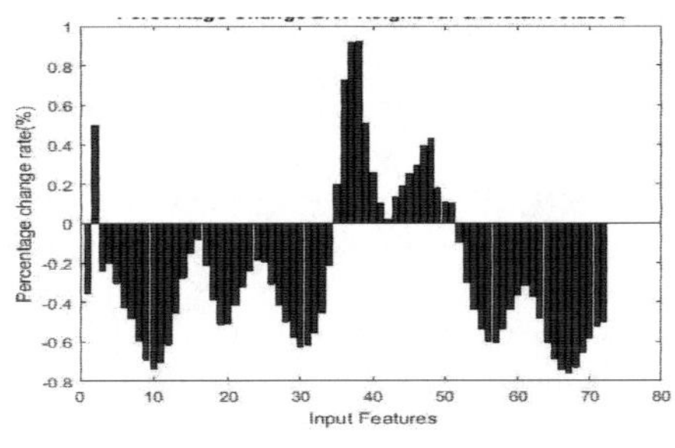

Figure 10: FWSOM's relevant inputs for class 2 - betaBand

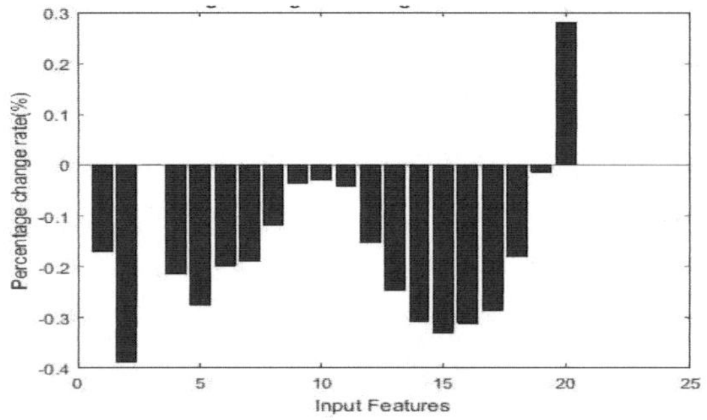

Figure 11: FWSOM's relevant inputs for class 1 - deltaBand

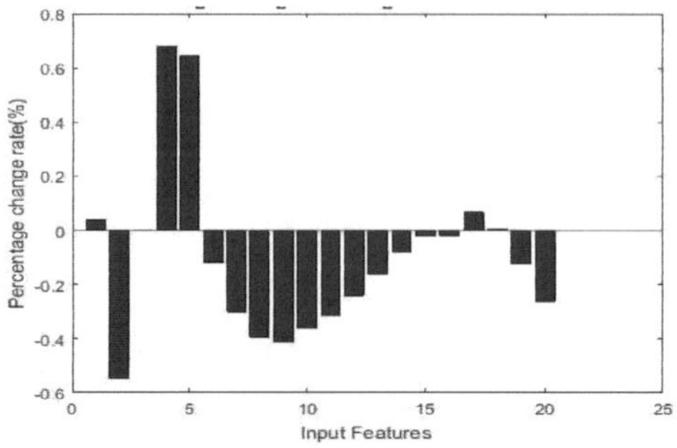

Figure 12: FWSOM's relevant inputs for class 2 - deltaBand

5. Conclusion

The Temporal nature of the EEG data makes it difficult for classifiers especially linear to separate the groups distinctly without pre-processing, transformation and extraction of features that are well known to clearly distinguish different mental tasks (which is quite difficult to achieve in practice). We have demonstrated how the powerful topology property of the SOM in conjunction with a novel feature weighting method can be used to improve the classification of this data.

The FWSOM in addition to achieving higher classification accuracy has given insights into what features are important in the classification of each class (left and right-hand movements), and these are corroborated by already published work in this area. The exact location of each of the input features identified by the FWSOM method as important is an interesting area to explore in future work extension of this paper.

The results also show that the identified features as relevant can also be used to improve the classification performance of other classification methods, which highlights again the importance of the features used during the SOM training process. The application of the FWSOM method has the potential to help identify relevant features in any given dataset and to give much improved classification accuracy over other methods.

Future work will focus on the usage of growing SOM methods so that a defined SOM lattice size is not required, a relaxing of the assumption for a class to be defined by a single relationship, and also improvements to the calculation of relevant and irrelevant inputs.

References

[1] ALOMARI, M.H., SAMAHA, A. AND ALKAMHA, K. 2013. Automated classification of L/R hand movement EEG signals using advanced feature extraction and machine learning. arXiv preprint arXiv:1312.2877

[2] ŞEN, B., PEKER, M., ÇAVUŞOĞLU, A. AND ÇELEBI, F.V. 2014. A comparative study on classification of sleep stage based on EEG signals using feature selection and classification algorithms. Journal of medical systems 38, 18.

[3] GARRETT, D., PETERSON, D.A., ANDERSON, C.W. AND THAUT, M.H. 2003. Comparison of linear, nonlinear, and feature selection methods for EEG signal classification. IEEE Transactions on neural systems and rehabilitation engineering 11, 141-144.

[4] AL GHAYAB, H.R., LI, Y., ABDULLA, S., DIYKH, M. AND WAN, X. 2016. Classification of epileptic EEG signals based on simple random sampling and sequential feature selection. Brain Informatics 3, 85-91.

[5] GOLDBERGER, A.L., AMARAL, L.A., GLASS, L., HAUSDORFF, J.M., IVANOV, P.C., MARK, R.G., MIETUS, J.E., MOODY, G.B., PENG, C.K. AND STANLEY, H.E. 2000. PhysioBank, PhysioToolkit, and PhysioNet: components of a new research resource for complex physiologic signals. Circulation 101, E215-20.

[6] KOHONEN, T. 1990. The self-organizing map. Proceedings of the IEEE 78, 1464-1480.

[7] FLETCHER, T. 2009. Support vector machines explained. Tutorial Paper.

IOP Conf. Series: Materials Science and Engineering **261** (2017) 012006 doi:10.1088/1757-899X/261/1/012006

[8] STEINWART, I. AND CHRISTMANN, A. 2008. Support vector machines. Springer Science & Business Media,

[9] SÖRNMO, L. AND LAGUNA, P. 2005. Bioelectrical signal processing in cardiac and neurological applications. Academic Press,

[10] VUCKOVIC, A. AND SEPULVEDA, F. 2008. Delta band contribution in cue based single trial classification of real and imaginary wrist movements. Medical & biological engineering & computing 46, 529-539.

[11] LI, Z., XU, J. AND ZHU, T. 2015. Recognition of brain waves of left and right hand movement imagery with portable electroencephalographs. arXiv preprint arXiv:1509.08257

[12] MURUGAPPAN, M. AND MURUGAPPAN, S. 2013. Human emotion recognition through short time Electroencephalogram (EEG) signals using Fast Fourier Transform (FFT). In Signal Processing and its Applications (CSPA), 2013 IEEE 9th International Colloquium on, Anonymous IEEE, 289-294.

[13] SOKOLOVA, M. AND LAPALME, G. 2009. A systematic analysis of performance measures for classification tasks. Information Processing & Management 45, 427-437.

[14] SLEIGHT, J., PILLAI, P. AND MOHAN, S. 2009. Classification of executed and imagined motor movement EEG signals. Ann Arbor: University of Michigan 1-10.

IOP Conf. Series: Materials Science and Engineering **261** (2017) 012007 doi:10.1088/1757-899X/261/1/012007

A Heuristic Bioinspired for 8-Piece Puzzle

M O Machado[1, a], P A Fabres[2, b] and J C L Melo[3, c]
[1]UFMS/CPAN - Brazil, [2]UFMS/CPAN - Brazil and [3]UFRJ - Brazil

[a]murilo.machado@ufms.br; [b]phelipe.fabres@ufms.br; [c]cabrallima@ufrj.br

Abstract. This paper investigates a mathematical model inspired by nature, and presents a Meta-Heuristic that is efficient in improving the performance of an informed search, when using strategy A * using a General Search Tree as data structure. The work hypothesis suggests that the investigated meta-heuristic is optimal in nature and may be promising in minimizing the computational resources required by an objective-based agent in solving high computational complexity problems (n-part puzzle) as well as In the optimization of objective functions for local search agents. The objective of this work is to describe qualitatively the characteristics and properties of the mathematical model investigated, correlating the main concepts of the A * function with the significant variables of the metaheuristic used. The article shows that the amount of memory required to perform this search when using the metaheuristic is less than using the A * function to evaluate the nodes of a general search tree for the eight-piece puzzle. It is concluded that the meta-heuristic must be parameterized according to the chosen heuristic and the level of the tree that contains the possible solutions to the chosen problem.

1. Introduction

Works that uses the 8-Piece Puzzle Problem to introduce artificial intelligence are quite common. [1][2][3][4][5]. Mainly, the ones with focus in optimizing the amount of memory spend and the processing timing required by the computer to solving Stochastic problems, The search strategy A* is a model well discussed in the literature. The different properties and qualities resulting from the chosen heuristic show us that, if it is permissible and consistent, the search is optimal [6]. However, depending on the problem, memory usage and runtime may be equivalent to NP-Complete problems [7].

Some papers concentrate their efforts on the search for extensions of the two classic heuristics: h1 (quantity of pieces out of place) and h2 (distance of Manhattan). Be investigating the search strategies A * and AIA * (which is the iterative version of A *) and presenting a third heuristic with small modifications in the distance of Manhattan. Or, showing significant improvements in memory required [1] [8].

Proper heuristics offer optimum results for these problems, especially in the areas of Artificial Intelligence (IA) and Optimization of Algorithms [9]. Computational Intelligence (CI) has some interesting approaches to develop computational models, such as Neural Networks, Fuzzy Logic and Genetic Algorithms. Indeed, IC has been widely used in computer simulations and has shown to be successful in many models of real problems [10].

An important improvement in this context was the integration of CI with Bio-Inspired Computing. In Swarm Intelligence (SI), the main objective is to understand how the interactions between agents occur, and how these agents behave in the face of changes in their environment [11] [12]. The central hypothesis is that there is intelligence in the overall behaviour of a group of agents, although each member of this group follows natural rules without knowing who controls or guides the group [13] [14].

Ant colony models [15] and school of fish [16] has obtained important solutions to problems that require high computational power. Mathematical models are powerful tools for understanding and describing the actual physical phenomena. In order to validate mathematical models, computational models are widely used to simulate these phenomena. A robust computational model usually deals with a nonlinear problem with solutions that do not have trivial analytical results.

The Problem of the n-Piece Puzzle is an excellent example to investigate the performance of search algorithms, since it presents an exponential complexity in relation to the quantity of pieces of the puzzle. Different search strategies have been applied in solving this problem, where the amount of computational memory and processing time required to it are the main variables to be investigated.

The general objective of this article is to present new properties of a mathematical model [8], being described qualitatively these new characteristics and correlating with the concepts of the search strategy A *.

Also, it is intended to solve the problem of the Eight-Piece Puzzle using the mathematical model presented and comparing the results obtained with the search strategy A *. The amount of memory used will be the parameter of comparison between the models investigated.

2. The meta-heuristic

The search strategy A * [6] presents the function $f(n) = g(n) + h(n)$, where the cumulative cost of the search is provided by $g(n)$, and $h(n)$ is the heuristic that When the search is close to the solution.

When a search tree is used as the data structure, the number of expanded nodes and number of nodes stored in the border variable are the computational variables related to the complexity of the problem. The literature shows that the strategy A * is adequate for solving the problem of Puzzle n parts when chosen a good heuristic[17]. There are two permissible and consistent heuristics widely used in the investigation of this problem, with $h1$ and $h2$.

According to Russell [10], the simplest way to reduce memory requirements is to use AIA *. However the same problems (such as generation of repeated nodes) from iterative deepening are encountered. The mathematical model presented here was built upon the investigation of the behaviour of bird migration [18]

$$L(\alpha, V_a) = \frac{m*V_a^2}{2} - m * V_a * V_w * cos(\alpha) + \frac{k*V_a^2}{2} \quad (1)$$

The migration of birds has some interesting peculiarities, such as the "same destination" in numerous migratory journeys made by new-born birds, endowed with a very efficient orientation and until today not very understood. Attempts to explain phenomena of this nature are generally grounded in energy optimization theories, which induce (in certain configurations) that birds tend to minimize their energy expenditure.

From this model, a meta-heuristic was generated based on the Newton-Raphson method [10], and we used the operator $\partial L/\partial V_a/\partial^2 L/\partial V_a^2$ of the Hessian matrix to find the points of (1) and then an adjustment was made to the value f obtained by the strategy A *, resulting in equation 2.

$$f(n) = f(n) - \partial L/\partial V_a/\partial^2 L/\partial V_a^2 \quad (2)$$

This paper intends to investigate the performance of the metaheuristic (equation 2) in relation to the search strategy A * when the 8-Piece Puzzle is solved. Note that the basis for solving the problem is the Tree General Search Algorithm and what differs is the evaluation of tree nodes by metaheuristic or A * strategy.

3. Experiment

The experiment consists in solving the problem of the 8-piece puzzle by taking 1000 samples of initial states, which were obtained after 20 random movements from the problem solution (objective state) [19]. Thus, the existence of solutions among the 20 First Levels of the Search Tree is guaranteed.

From the meta-heuristic, two mathematical functions were generated to be applied in the experiment: L1, which is equation 2 with h1 inserted in the domain and L2, which is equation 2 with h2 inserted in the domain.

In the meta-heuristic, the domain of the generalized coordinates α and V_a will be parameterized according to the limits of the heuristic used, and the values of referring to variable g corresponding to the model A *, then the limits will be adjusted within the appropriate limit for the Problem investigated. In this way $h1$ can range from 0 to 8, and $h2$ can range from 0 to 26. The domain of the generalized coordinate α limited from 0 to 180 o will be divided by 8 linearly spaced points when used $h1$ and divided by 26 linearly spaced points when it traverses to $h2$. The generalized coordinate domain V_a of 0 to 40 m / s will be divided between 8 linearly spaced points when used $h1$ and divided by 26 linearly spaced points when $h2$ is used.

Samples were provided as input to the Tree General Search Algorithm using $h1$ and $h2$ (using the A * search strategy), and $L1$ e $L2$ (using metaheuristics). The tree is generated from the initial node, entered as input into the algorithm, and consists of the nodes results of the expansion and are stored on the edge, which is a sub tree with all possible states for the problem at that time.

As a result, we evaluated the evaluation of each node expanded and stored on the edge [10], then the average of the total amount of expanded nodes is calculated to solve the problem for each example.

In the evaluation of the obtained results, a table was constructed to present the average of the generated nodes when using each heuristic and distinct heuristic goal, and a figure to show the evaluation of the state spaces to each expanded node in the course of a search.

4. Results

Figure 1 shows the behaviour of the evaluation functions $f(n)$ (figure to the left) and L figure to the right) used in this article. The image of the evaluation function $f(n)$ is contained in a discrete and limited universe between 0 and 60. The image of the evaluation function L is contained in a continuous universe with the image limited between -10 and 40.

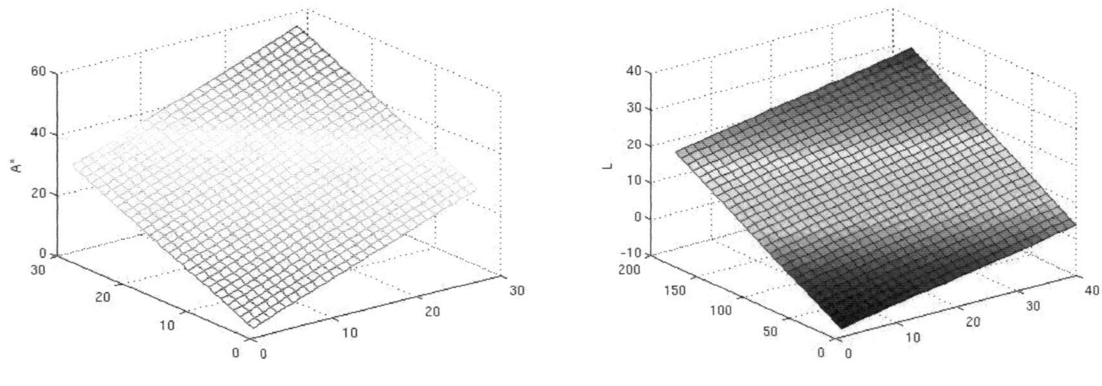

Figure 1. Figure 1 (a) shows the value of the function f (n) and Figure 1 (b) shows the value of the function L.

It is also observed that the voids contained in Figure 1 (a) are contemplated by Figure 1 (b) generating some zones of local maxima and minima, but without damaging the global growth presented by the function $f(n)$. It can also be observed that the evaluation of L does not have linear growth as the function $f(n)$. The grid on the evaluation surface of figure 1 (b) corresponds to the values obtained in equation 2 when the values of $g(n)$ and $h(n)$ are inserted. The values assigned in L to the constants m and k, and in how many parts and divided the domain of α and V_a are determinant to move this grid by the evaluation surface.

Figure 2 shows the evolution of the search space to each loop of the algorithm, showing the evaluation of each node that is inserted in the edge. The blue dots represent the upward growth of $f(n)$ and the green dots, referring to L, show some zones of minimum and maximum locations.

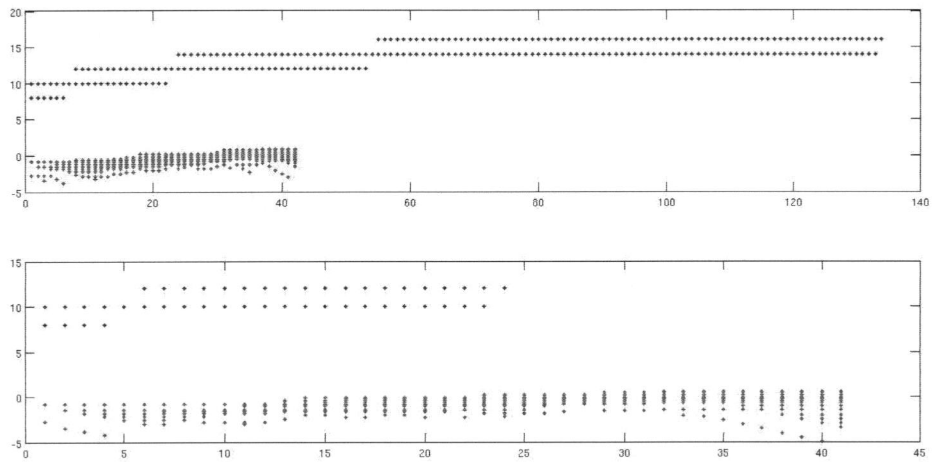

Figure 2. Figures 2 (a) and 2 (b) show the nodes contained in the space by the strategy A * and the nodes contained in the space with the Meta-Heuristic L, respectively.

The article focused on empirically investigating which better division of the domain of equation 2. Initially, the domain V_a was divided into 29 parts and the domain α into 26 parts and it was verified that the domain obtained good results when the value of the domain of V_a was decreased. Since h can not be greater than 26 and the sample used does not contain problems in depth greater than 20, there is a small probability of $h(n)$ being greater than 20. Following this hypothesis, the space of V_a was divided into smaller parts, and the results presented in table 1 (division into 18 parts) were verified.

It can be observed in figure 2 that the algorithm searches for the minimum of the function always choosing the node with the lowest evaluation of f or L. Since $f(n) > 0$ for all $g(n)$ and $h(n)$ the curve is ascending for $f(n)$ and descending to L since its nodes have negative evaluation in this domain. Figure 2 (a) shows 134 executions of $f(n)$ and 42 of L. Figure 2 (b) shows 24 executions of $f(n)$ and 41 of L Note that some nodes are overlapped by having same evaluation. The evaluation of the nodes by the function L is able to differentiate the nodes that had the same evaluation by the function $f(n)$. This can be observed in figure 2 (b) (second iteration). Note that the execution of the function $f(n)$ evaluates two nodes with the same value and the execution of the function L differentiates these same nodes with 3 different values.

Table 1 shows that the meta-heuristics $L1$ and $L2$ improve the performance of the heuristics $h1$ and $h2$ reducing the amount of expanded nodes. Note that $L1$ finds fewer solutions than $h1$ for solutions in levels 7 and 8. In contrast, $L2$ finds more solutions than $h2$ for that same level. It can be observed that both $L1$ and $L2$ find solutions at deeper levels than the heuristics $h1$ and $h2$. It can be said that the amount of memory required is smaller when recoding the evaluation function L to solve a set of random examples.

Table 1. Average amount of expanded nodes with quantity of solutions at different levels of the Tree. Columns 1 and 2 correspond to the results obtained when the A * strategy is applied using the heuristic h1 and the heuristic L1, respectively. Columns 3 and 4 correspond to the results obtained when using the strategy A * using the heuristic h2 and the heuristic L2, respectively

Level	h1	L1	h2	L2
1-2	3 (119)	3 (119)	3 (119)	3 (119)
3-4	5.3054 (203)	5.0936 (203)	5 (203)	5.0 (203)
5-6	10.1255 (247)	9.0810 (247)	7.9555 (247)	7.5547 (247)
7-8	20.2105 (190)	16.2514 (183)	12.4324 (185)	10.2474 (190)
9-10	45.7760 (125)	36.1846 (130)	20.4524 (126)	16.0968 (124)
11-12	113.2812 (64)	81.2308 (65)	38.0597 (67)	31.3607 (61)
13-14	299.5217 (23)	224.9545 (22)	81.2500 (24)	90.0385 (26)
15-16		310.75 (4)	81 (3)	112,3333 (3)
17-18				0 (0)
19-20				62 (1)

The metaheuristic shows promise to complement or replace the heuristic h without impairing its properties of admissibility, as well as the generalized coordinate α (Angle of the direction of the wind with the bird) is the main cost that the bird has in its migratory journey, And has an implicit analogy with the variable g presented by the search strategy A*.

References

[1] Nayak D and B Majhi 2014 *Analysis and Implementation of Admissible Heuristics in 8 Puzzle Problem* (Department of Computer Science and Engineering National Institute of Technology Rourkela Rourkela).
[2] Tuccar M 2012 Analyzing the A* Search Heuristics with the Solutions of the 8 (University of Massachusetts Lowell).
[3] Paderborn A R 1993 Complete Solution of the Eight-Puzzle and the Bene t of Node Ordering in IDA* (Center for Parallel Computing Warburger, Paderborn, Germany).
[4] Hansson O and Mayer A and Yung M 1992 Criticizing Solutions to Relaxed Models Yields Powerful Admissible Heuristics (Elsevier Science Publishing Co).
[5] Demaine D E and Rudoy M 2017 A simple proof that the (n^2 -1) puzze is hard (MIT Computer Science and Artificial Intelligence Laboratory).
[6] Hart P and Nils J and Raphael B 1968 A Formal Basis for the Heuristic Determination of Minimum Cost (IEEE Transactions On Systems Science and Cybernetics).
[7] Wilt C and Ruml W 2015 Building a Heuristic for Greedy Search (AAAI Publications, Eighth Annual Symposium on Combinatorial Search).
[8] Bauer B 1994 The Manhattan Pair Distance Heuristic for the 15-Puzzle (Universitat-GH Paderborn, Paderborn, Warburger).
[9] Michalewicz Z and Fogel D B 2004 How to solve it: Modern heuristics (University of Massachusetts Lowell).
[10] Engelbrecht A P 2002 Computational intelligence an introduction (Center for Parallel Computing Warburger, Paderborn, Germany).
[11] Beni G and Wang J 1998 Swarm intelligence in cellular robotic systems, proceedings of the nato advanced workshop on robots and biological systems (Elsevier Science Publishing Co).

[12] Bonabeau E and Dorigo M and Theraulaz G 1999 Swarm intelligence, from natural to artificial systems (IEEE Transactions On Systems Science and Cybernetics).
[13] Beekman M and Sword G A and Simpson S J 2008 Biological foundations of swarm intelligence (AAAI Publications, Eighth Annual Symposium on Combinatorial Search).
[14] Nicoud, J and D Deneubourg, J and Floreano D 2009 Swarm intelligence in autonomous collective robotics: From tool to the analysis synthesis of distributed control strategies (Universitat-GH Paderborn, Paderborn, Warburger).
[15] Pat A and Hota A R 2011 Ant colony optimization and hypergraph covering problems (University of Massachusetts Lowell).
[16] Kobilarov M and Crane K and Desbrun M 2009 Lie group integrators for animation and control of vehicles (Center for Parallel Computing Warburger, Paderborn, Germany).
[17] Machado O M and Lima C and Gutierrez L B and Jimenez F S 2016 Uma Meta-Heurística Bio-Inspirada para um Agente de Busca.
[18] Machado O M and Lima C 2012 Predição de ajustes em vôos de aves migratórias sazonais noturnas: um simulador com adequação nebulosa (UFRJ- Rio de Janeiro- RJ).
[19] Russell and Stuart J and Peter Norvig 1995 Artificial Intelligence: A Modern Approach (Englewood Cliffs, NJ: Prentice Hall).

IOP Conf. Series: Materials Science and Engineering **261** (2017) 012008 doi:10.1088/1757-899X/261/1/012008

Chinese Sentence Classification Based on Convolutional Neural Network

Chengwei Gu[1], Ming Wu[2] and Chuang Zhang[3]

[1, 2, 3]School of Information and Communication Engineering, Beijing University of Posts and Telecommunications
Xitucheng Road No.10, Beijing, P.R.China

[1]gcw831tjs@163.com, {[2]wuming, [3]zhangchuang}@bupt.edu.cn

Abstract. Sentence classification is one of the significant issues in Natural Language Processing (NLP). Feature extraction is often regarded as the key point for natural language processing. Traditional ways based on machine learning can not take high level features into consideration, such as Naive Bayesian Model. The neural network for sentence classification can make use of contextual information to achieve greater results in sentence classification tasks. In this paper, we focus on classifying Chinese sentences. And the most important is that we post a novel architecture of Convolutional Neural Network (CNN) to apply on Chinese sentence classification. In particular, most of the previous methods often use softmax classifier for prediction, we embed a linear support vector machine to substitute softmax in the deep neural network model, minimizing a margin-based loss to get a better result. And we use tanh as an activation function, instead of ReLU. The CNN model improve the result of Chinese sentence classification tasks. Experimental results on the Chinese news title database validate the effectiveness of our model.

1. Introduction

Sentence classification is the fundamental issue in NLP. Solving this problem opens many doors in NLP, such as information retrieval, machine translation and automatic digest. Since Sentence classification is such significant in NLP, there exists a lot of previous work in this field. The traditional sentence classification method is mainly based on the principle of statistics, using machine learning model. Using manually annotated data sets for supervised training to obtain classifiers, and then classify the new data. Among them, the general learning methods include Naive Bayes Classifier, Maximum Entropy Model, Logistic Regression, and so on. In the semantic classification task, the sentences have the following characteristics: (I) Single sentence language composition is small: each sentence contains a few words, up to dozens of words, some sentences do not even include the complete sentence composition, often pile the words; (II) The emergence of new words for the existing classifier is very difficult to deal with the unknown words (Out of Vocabulary). These characteristics lead to the traditional sentence classification method need to update the dictionary for the sentence data set and features, but this method is a drop in the bucket.

In this paper, we concentrate on classifying Chinese sentences by deep neural network. After a couple of pioneer works [1][2], neural network is showing to be effective in the research and applied to many NLP tasks, such as semantic parsing [3], translation task [4]. In the field of NLP, many deep learning models have been applied to the study of word embedding, and using word embedding to accomplish a couple of tasks. The word embedding is a sparse V-dimensional (V is the size of the vocabulary) vector mapped to low-dimensional vector through the hidden layer, and each dimension can represent some semantic information. Neural network model was applied to computer vision research achieved excellent results [5], and then obtain good results in sentiment analysis, sentence retrieval, sentence model [6] and other NLP tasks. As a complicated architecture with huge number of parameters, it is rather important to have a careful parameter adjust together with a large scale of training samples. More specifically, we use the Chinese news title database to validate this deep neural model. Experimental results on the Chinese news title database confirm a performance improvement.

Recently, the success of the deep learning in the field of the NLP has attracted more attention. In contrast to the traditional machine learning methods, sentence representations learnt by deep network structure have shown their great potential in various NLP tasks. There are many models in deep learning fields, one of the greatest breakthroughs in sentence classification is deep CNN model. CNN has achieved good results in classifying corpus for short sentences. In this paper, we propose a novel and different CNN to extract high level sentence representations for classification. Compared to traditional models [7], we adjust the network architecture by accomplishing a linear support vector machine to substitute softmax, and optimizing the parameters by minimizing a margin loss. We also alter the activation function and adjust the neural network architecture. And in this paper the task we completed is the Chinese sentence classification, so we also need to segment the Chinese sentence. These changes improved deep learning model not only making up the shortfall in traditional machine learning representation extraction, but also gaining better performance in Chinese sentence classification.

2. Related work

Sentence classification has received great deal of attention in the last few decades [1] [7]. So there is a lot of research on sentiment analysis, or more generally on sentence classification tasks. Traditional methods always have two stage that one is extraction of hand-crafted features followed by a classifier. Typical features include Document Frequency [8], Information Gain, Cross Entropy, TF-IDF [9] and N-grams [10]. The appropriate features take an important role in accomplishing the final algorithm. And then algorithms take the major in training and testing work. So the extraction of features becomes the most important and most challenging work in sentence classification. Choose a good feature is half of the battle. These solutions are defeated by the deep learning models recently. We use Chinese corpora which applied by Sogou Labs for our experiments. At first, we use One-hot Representation to represent a word, the vector size is equal to the the number of vocabulary which can be large. Then, words have been mapped into a low-dimensional space using word2vec [11] which was proposed by Google. And then these word embedding has low vector size which then be put into the classifier. These solutions preform not good, because all words order is neglected.

Approaches that use CNN followed by a classical classifier have been proposed recently. In other words, the CNN herein is to extract features, to obtain good and high level sentence representations. A fairly shallow CNN model has been proposed by Kim [7]. This neural network architecture is rather simple. This model use word2vec toolkit to get word embedding as input. And then one convolutional layer followed by a max pooling layer. In this model, three filters of different sizes are used to complete the convolution. Then followed by a fully connected layer using drop-out to avoid over fitting. A similar model was proposed by Kalchbrenner [12]. A significant difference between these two models is that this model has many temporal k-max pooling layers and much deeper than previous models. This change makes it possible to get the k most vital representations in a sentence regardless of their position. Inspired by the previous work, we proposed a new and novel CNN model with SVM as classifier served as a supervised learning solution.

3. Model Architecture

The CNN model that we proposed is used for Chinese sentence classification. Our design for this model architecture has taken the risk of over fitting into account. The overall architecture of our model is shown in Figure 1, the neural network consists of word embedding as input layer, two convolutional layers, tanh as activation function followed by two fully-connected layers with a small amount of neurons and SVM layer at last.

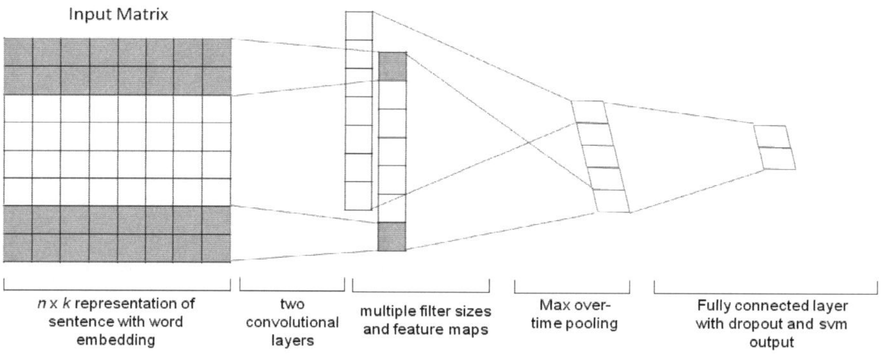

Figure 1. Model Architecture

3.1 Chinese Word Segmentation

In this paper, we use jieba-analysis toolkit to segment the SougouT corpus. We obtain the words after segmenting sentences. And then these words were put into the neural language model to get the word embedding.

3.2 Pre-trained Word Embedding

We initialize our word embedding with that acquired from an unsupervised neural language model. The model is a popular solution to improve performance under the condition that we do not have a very large supervised training set. We use the available word2vec toolkit which proposed by Google [11]. We use the SogouT corpus data set to train the word2vec toolkit to obtained the word embedding. We use the continuous bag-of-words architecture to obtain these vectors. Words that are not in the pre-trained words set are initialized randomly. The word embedding can be fine-tuned in our model. Word embedding in the paper we use are static and non-static. In our model architecture, showed in Figure 1, each filter is applied to two channels that one is static, another is fine-tuned. And then results are added to computed convolutional layer output.

3.3 CNN Architecture

We obtain the word embedding which is the k-dimensional word vector from the previous step. A sentence of length n can be represented as a $n * k$ matrix

$$x_{1:n} = x_1 * x_2 * x_3 \ldots * x_n \quad (1)$$

which * is the cascade operator. Generally, we use $x_{1:n}$ represent as cascading of words x_1, x_2, x_3... x_n, in other words sentence of length n. Then followed by a convolution layer. We use different size of filters to perform convolution on the input matrix. The filter $w \in R^{lk}$, which is applied to a window of length words to obtain a high level representation. For example, we use filter of length l to generate representation by

$$c_i = f(w \cdot x_{i:i+l-1} + b) \quad (2)$$

in the equation (2) b is a bias added to the convolutional layer out and f is a non-linear function called activation function. We use tanh to substitute ReLU in our model, because we acquired the promotion of result in the experiment. The filter is applied to produce a feature map

$$c = \{c_1, c_2, c_3, \ldots, c_{n-l+1}\} \quad (3)$$

with $c \in R^{n-l+1}$.

Most of previous convolution neural network model is fairly shallow which used to implement NLP task. This is due to the convolution of the token's n-gram feature, and the need for different n-gram length to model short and long-span relationship. In this paper, we then apply another convolution layer and a batch norm layer [13] and tanh activation function showed in Figure 2 in order to extract feature map. Next, the model followed by a max-pooling layer on the feature map and obtain the maximum value in the feature map as the representation corresponding to this size of filter. We use this max-pooling layer to to get the most essential feature in each feature map. The max-pooling operation can solve the different length of sentences problem, after this operation can make the output vector size be the size of the number of the filter. We extract one feature from one filter in this novel convolution neural network model. The model uses many filters with different slide window sizes to get a number of features. These features are then put into fully-connected layer, and then the output is fed to a SVM layer to complete training a neural network for classification. layer weights are updated via back propagating the gradients from the SVM layer. We use SVM layer substitute softmax to accomplish classification. The linear SVM layer is widely used for classification.

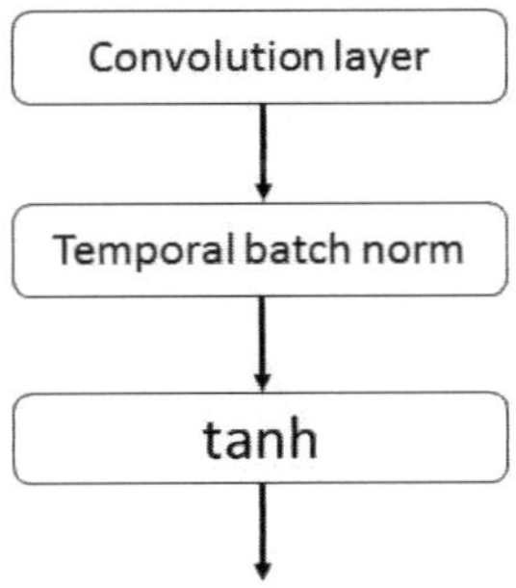

Figure 2. Convolutional layer Architecture

3.4 Regularization

We used dropout on the second last layer with a l2-nomalize constraint of the weight matrix. Dropout prevents over fitting via dropping out some connection randomly. That is, the second last layer output $= \{c_1, c_2, c_3, \dots, c_m\}$, instead of using

$$y = w \cdot z + b \tag{4}$$

for out y in the training step, dropout uses

$$y = w \cdot (z \; or \; 0) + b \tag{5}$$

Gradients are back propagated only via the units which are not dropout. At the test, we do not use dropout (not discard any of the network parameters to, then dropout layer output equivalent to input). Then, the learned weight vectors are scaled by p such that $w = p \cdot w$, which p is Bernoulli probability.

4. Experimental Results

The deep neural network model we proposed in the previous chapter was compared with the classical CNN approach and machine learning model on Chinese sentence classification. And in the experiment tensorflow [14] was used to implement our model on the machine with Titan X GPU.

4.1 Dataset and Experimental Setup

We choose SogouT corpus to train neural language model via word2vec toolkit. Then we obtain word embedding as input. We choose Chinese news title corpus as training data and testing data. One news title data set 9k (NT-9k): This data set contains 9000 sets of training data, 1000 sets of cross validation data, and 4000 sets of test data. The data are divided into nine categories: domestic, international, social, financial, stock, technology, sports, military and entertainment. The maximum length of the sentence is 21. The entire data set contains 140195 words (including the placeholder for the empty word). Another news title data set 16k (NT.16k): This data set contains 16,000 sets of training data, 1000 sets of cross validation data, and 4000 sets of test data.

4.2 Static Input vs. Fine-tune Input

We had expected that the fine-tune input would prevent over fitting (by ensuring that the learned vectors do not deviate too far from the original values) and thus work better than the static model, especially on smaller datasets. The results, however, are mixed, and further work on regularizing the fine-tuning process is warranted. For instance, instead of using an additional channel for the non-static portion, one could maintain a single channel but employ extra dimensions that are allowed to be modified during training. The part of classification results shows in Table 1.

Table 1. Part of classification results

	Domestic	International	Social	Financial	Stock	Technology
Static	0.842	0.874	0.811	0.823	0.812	0.857
Fine-tune	0.861	0.898	0.892	0.867	0.873	0.884

4.3 L2-SVM vs. Softmax

We compared performance among Naive Bayesian Model, the convolution neural network model with our novel model using SVM layer rather than softmax. All of models are tested using cross validation with the architecture we have introduced.

We may have a look at the validation curve of the softmax vs L2-SVM with the function of weight updates in Figure 3. When it came to the latter half stage of training process, weights were updated slower because learning rate get lowered. But our model with L2-SVM still had a little advantage with a lower averaged error rate.

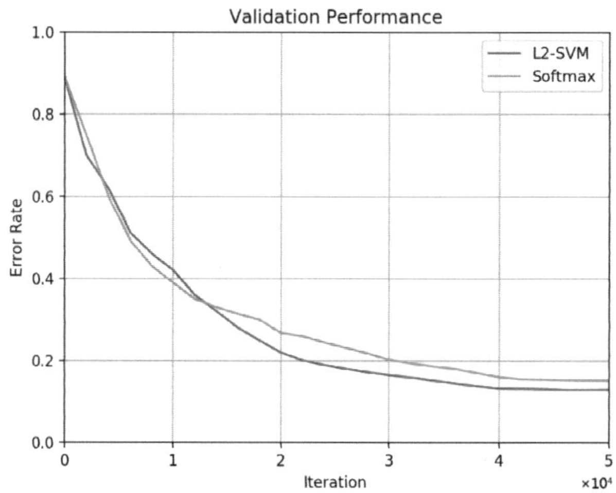

Figure 3. Validation performance between Softmax L2-SVM

4.4 Result

In order to validate our neural network model, we first compared it with traditional solution using traditional features have been mentioned previously, and next we compared our model with convolution neural network that is proposed by Yoon Kim[7], Different methods with its average accuracy on all category is shown as Table 2. Our model has a little promotion than other models. We set batch-size 64 through our experiment.

Table 2. Results of our CNN models against other methods

Models	Accuracy
Naive Bayesian Model	73.6%
CNN(Yoon Kim)	84.3%
CNN(Our Model)	87.1%

5. Conclusion

In the present work, we have proposed a new convolution neural network model architecture built on top of word2vec for Chinese sentence classification where a support vector machine was embedded to get learn a high-level hierarchical representation of a sentence. We train and test our model on the Chinese news title corpus and an experiment demonstrated the promotion that our model achieved. Our model prove that the input word embedding is effective to Chinese sentence classification. Although sentence abide by human rules and images are the original signals from the nature, short sentences and images have the same properties. In this paper, we concentrate on using convolutional neural network to accomplish sentence classification task. Applying similar model to other natural language processing tasks, in particular sentiment analysis is left for future research.

References

[1] Y. Bengio, R. Ducharme, P. Vincent. 2003. Neural Probabilitistic Language Model. Journal of Ma- chine Learning Research 3:1137–1155.
[2] Collobert R, Weston J. A unified architecture for natural language processing:deep neural networks with multitask learning[C]// International Conference. DBLP, 2008:160-167.
[3] Yih W T, He X, Meek C. Semantic Parsing for Single-Relation Question Answering[C]// Meeting of the Association for Computational Linguistics. 2014:643-648.
[4] Collobert R, Weston J, Karlen M, et al. Natural Language Processing (Almost) from Scratch[J]. Journal of Machine Learning Research, 2011, 12(1):2493-2537.
[5] Hinton G E, Srivastava N, Krizhevsky A, et al. Improving neural networks by preventing co-adaptation of feature detectors[J]. Computer Science, 2012, 3(4):págs. 212-223.
[6] Kalchbrenner N, Grefenstette E, Blunsom P. A Convolutional Neural Network for Modelling Sentences[J]. Eprint Arxiv, 2014, 1.
[7] Kim Y. Convolutional neural networks for sentence classification[J]. arXiv preprint arXiv:1408.5882, 2014.
[8] Robertson S. Understanding inverse document frequency: on theoretical arguments for IDF[J]. Journal of Documentation, 2004, 60(5):503-520.
[9] Joachims T. A Probabilistic Analysis of the Rocchio Algorithm with TFIDF for Text Categorization[C]// Fourteenth International Conference on Machine Learning. Morgan Kaufmann Publishers Inc. 1997:143-151.
[10] Damashek M. Gauging Similarity with n-Grams: Language-Independent Categorization of Text[J]. Science, 1995, 267(5199):843-8.
[11] Goldberg Y, Levy O. word2vec Explained: deriving Mikolov et al.'s negative-sampling word-embedding method[J]. Eprint Arxiv, 2014.
[12] Conneau A, Schwenk H, Barrault L, et al. Very deep convolutional networks for text classification[J]. arXiv preprint arXiv:1606.01781, 2016.
[13] Sergey Ioffe and Christian Szegedy. 2015. Batch normalization: Accelerating deep network training by reducing internal covariate shift. In ICML, pages 448–456, Lille, France.
[14] Abadi M, Agarwal A, Barham P, et al. TensorFlow: Large-Scale Machine Learning on Heterogeneous Distributed Systems[J]. 2015.

IOP Conf. Series: Materials Science and Engineering **261** (2017) 012009 doi:10.1088/1757-899X/261/1/012009

Auto-Associative Recurrent Neural Networks and Long Term Dependencies in Novelty Detection for Audio Surveillance Applications

A Rossi[1], F Montefoschi[1], A Rizzo[1], M Diligenti[2] and C Festucci[3]

[1] Department of Social, Political and Cognitive Sciences, University of Siena, Italy
[2] Department of Information Engineering and Mathematics, University of Siena, Italy
[3] Banca Monte dei Paschi di Siena, Siena, Italy

Email: alessandro.rossi2@unisi.it

Abstract. Machine Learning applied to Automatic Audio Surveillance has been attracting increasing attention in recent years. In spite of several investigations based on a large number of different approaches, little attention had been paid to the environmental temporal evolution of the input signal. In this work, we propose an exploration in this direction comparing the temporal correlations extracted at the feature level with the one learned by a representational structure. To this aim we analysed the prediction performances of a Recurrent Neural Network architecture varying the length of the processed input sequence and the size of the time window used in the feature extraction. Results corroborated the hypothesis that sequential models work better when dealing with data characterized by temporal order. However, so far the optimization of the temporal dimension remains an open issue.

1. Introduction

Automatic Sound Recognition (ASR) can be subdivided into two main categories: *speech* and *non-speech* problems. In the Machine Learning literature the term ASR is usually referred to the second area, whereas the first one is a separated field of research that has been investigated in-depth in the last years. However, also non-speech ASR has attracted increasing and wide ranging interest in recent years [1]. Many applications have been investigated like, for example, Music Information Retrieval [2], environmental sound event recognition [3][4] and Audio Surveillance (AS)[5], which is the main subject of investigation of this paper. Restricting our attention to classification problems, these are divided into two major categories: when the classification aims at detect different events from a categorization decided in advance [6][7], or when the task is to distinguish normal events from abnormal ones. The second assumption seems more general in typical surveillance and system monitoring scenarios, and it is usually referred to as *Novelty Detection* (ND) [8]. The general idea is to model a function $z(x)$ by training a learning agent on data considered to be normal, for which a large number of examples is usually available. The function $z(x)$ can be interpreted as a *novelty score* and the task is often defined as a one-class classification problem by assigning to the abnormal class those samples x for which $z(x) \geq k$, where k is a fixed threshold. From a theoretical point of view, this scenario is related to the statistical problem of outlier detection [9], trying to model a 1-class data distribution to detect possible anomalies or to remove possible noised data in general circumstances [10][11]. Different techniques have been investigated, employing different classifiers and possible sets of features [8]. Despite the intrinsic temporal nature of the studied environment, many proposed solutions neglect to represent the input as a sequence of frames, instead merging the input frames into a single static data representation and then using classical non-recursive machine learning approaches to perform learning and inference. For example, Gaussian Mixture Models are employed in [6][11], whereas Support Vector Machines are used in [12]. As a matter of fact, only in a few cases generative models processing temporal sequences in the form of Hidden Markov Models had been used [13].

Recent successful applications of Recurrent Neural Networks (RNNs) based on Long-Short Term Memory (LSTM) cells to sequences modeling [14][15][16] [17] suggests that the same advancements could be obtained in the field of Novelty Detection for AS. This has been pioneered in [18], where remarkable improvements in prediction performances by using unsupervised sequence-to-sequence models. The exploitation of RNNs seems to be sound in this context, since it is difficult to fix in advance the size of the considered temporal window during the processing of the audio signal. In particular, the capability of LSTM-RNNs to automatically learn the temporal correlations in the input signal could represent a crucial advantage in the considered application.

These ideas are explored in this work by the analysis of the behavior of a general sequence-to-sequence model used to process an audio signal in an unsupervised way. As already said, we assume to have not any a-priori knowledge on possible abnormal events during the training, when only the audio describing the environment during normal daily activity is available. A first LSTM module is embedded in an Auto-Encoder architecture and used to encode each sequence to a compressed feature representation. The obtained vector is then decoded by a similar module trained to reproduce the original input. The training phase indeed consists into minimizing the reconstruction error comparing the generated output with the corresponding input on normal data samples. Hence, the trained model is used to make predictions on new data in presence of various kinds of anomalies. A potential degradation in the reconstruction quality is used to detect abnormal events and it is possible to analyze how the model can learn when varying the temporal scope of its memory. A larger temporal memory is obtained by processing longer input sequences, allowing the network to exploit correlations that are farther away in time. In particular, this should help in presence of structured events, in which the unusual composition of singular occurrences, each one similar to normal ones, could represent a potential dangerous situation. The disadvantage of processing longer input sequences is the increased computational time and the more limited generalization capabilities of the network, as reconstructing long sequences is a much harder learning task.

After the description of the used architecture, this paper introduces the experimental setup and it shows how processing sequences of input frames, instead of static single data representations, affects the classification results. Interestingly, the proposed dynamic model is shown to outperform the static one even when the input data spans over the same temporal window, showing the importance of proper sequence modeling for this task.

2. Sequence-to-sequence Audio signal modelling

RNNs represent a powerful neural architecture recursively processing an input sequence by carrying on a state representation, based on the previously observed inputs. At each single step, the computation is performed using a standard Multi-Layer Perceptron. Back-Propagation through Time (BPTT) [19] is a training algorithm that extends the classical back-propagation procedures to this class of networks. Unfortunately, this training algorithm suffer of the well-known problem of vanishing gradients [20], making it difficult to learn correlation dependencies when processing long input sequences [21]. Long-Short term memory cells had been introduced [17] to enforce the error flow through different temporal states of the network by the definition of special units, called *gates*, capable of transmitting information within several time steps, while also maintaining the response to any local information. These architectures have been first proposed two decades ago, but have not been fully exploited for many years, because of the computational requirements and the difficulties in effectively training them. Recent advancements on computer architectures and on deep networks training have allowed to train LSTM networks effectively even on a large amount of training data. LSTM-RNNs have been shown to outperform state-of-the-art algorithms especially in problems that can be naturally represented via a sequence-to-sequence model, as for example in Handwriting Recognition [14] and Natural Language Processing [15][16]. The basic sequence-to-sequence approach, introduced and now widely popular in Machine Translation [22][23][24], consists in the use of a LSTM cell to encode an input sequence of variable length into a vector of numbers. Hence, a dual LSTM cell can be used to generate the output sequence according to the prediction to be performed, starting from the vector of numbers encoding the input. Due to the generality of this Encoding-Decoding procedure, the extension to several stacked layers is straightforward. For the same reasons, it can be easily adapted to our problem, due to the nature of the audio signal as a continuous stream of data. Indeed, assuming to represent an input frame by a feature vector, the audio signal is represented by a sequence of vectors. Hence, the network is trained to reproduce the sequence itself by minimizing the reconstruction error when processing the sequence of input vectors. This model, performing ND for AS tasks, has been already proposed in [18], showing to outperform classic approaches. A visual representation of the architecture is depicted in figure 1.

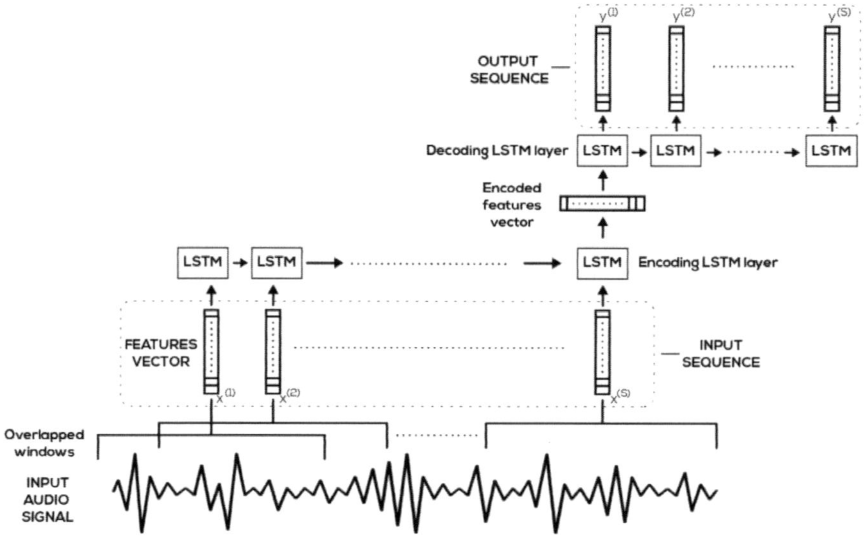

Figure 1. Global description of the Auto-Encoder architecture composed by Encoding and Decoding LSTM layers to compress and reconstruct the provided feature representation of input sequences.

The aim of this work is to analyse the relation between the quality of the information extracted at a feature level with the one learned by the RNNs. In particular, we start by extracting a feature representation $x^{(i)}$ of the audio signal, which will be explained in more details in the following of his section. This representation is computed on a window w composed by a certain number of samples from the raw audio signal, expressing a specific characterization of the input over a given interval. The width w of this window is fixed in advance, and so is the amount of information embodied by the generated representation. Hence, a sequence of length s of these feature vectors is assembled in order to be fed to the learner, covering a certain time interval I. The obtained interval I is proportional to the window size w and to the length of the sequence s. In particular, if we pose $r = 1 - h$, where h is the rate of overlap between two consecutive feature windows, the input size I spans over a temporal interval equal to: $I = w + (s-1) \cdot rw = w \cdot [1 + r(s-1)]$. A sketch of how the signal is fed into the processing architecture is depicted in figure 1. A first part of the experimental analysis consists in comparing the performances obtained by increasing s while keeping w fixed. Moreover, we also propose a comparison on different configurations by covering the global input window I in different ways while maintaining the rate w/s constant. A remark has to be done on the case $s = 1$, since in this case the RNNs architecture collapses into a standard Auto-Encoder (AE) [25] with one hidden layer to perform dimensionality reduction, i.e. we have:

$$hu < |x| = |y| \quad (1)$$

where hu is the number of hidden units and $|x| = |y|$ are the dimension of both the input and the (reconstructed) output vectors. The size of the hidden layer is chosen to be smaller than the size of the original input to prevent the network from learning trivial configurations.

3. Experiments

As already said, we focus on the exploration of the correlations of temporal intervals of different size and type. Since many investigations on the best features for ASR task can be found in the literature 0, we avoid further explorations on this part, focusing on the behaviour of the representational agent w.r.t. to a fixed input representation. The features are extracted exploiting the *librosa* package [26] by computing for each window of w samples the Zero-Crossing Rate (ZCR), 20 Mel-frequency cepstral coefficients (MFCCs) and their 20 derivatives (computed by the first order method), since similar configurations are proved 0[6][27] to embody a good general representation in many ASR task, both in speech and non-speech environments. Each extracted feature is then normalized to null mean and standard deviation equal to one. The window size w has been varied between 4096 and 32798 samples as reported in table 1, while keeping $r = 0.25$ fixed. In table 1 we reported the real-time window I obtained by varying the window size w and the length s of the sequence processed by the encoding LSTM module. As already said, $s = 1$ indicates that the input feature vector is processed by a static Auto-Encoder. In this case, since the decreasing rate of s (0.2 instead of 0.5) is different w.r.t. the ones among the other sequences, the w is adapted to obtain a coherent window in term of real world time.

Table 1. Real time interval length (in seconds) corresponding to the input window size analysed in each configuration, composing the feature extraction part and the sequence computed by the agent.

Input Window Size I (seconds)				
Sequence Length s (input steps)	Window Size w (samples)			
	4096	**8192**	**16384**	**32768**
40	3.71	7.43	14.86	29.72
20	1.85	3.71	7.43	14.86
10	0.92	1.85	3.71	7.43
5	0.46	0.92	1.85	3.71
1	0.09	0.18	0.46	0.92

Because of the difficulty in finding publicly available datasets, audio tracks (sampled at 44100 kHz)have been generated by recording a week plus one day of daily activity of the help desk of the library of the Engineering Department at the University of Siena in the time interval 8AM – 8PM. An environmental microphone was placed in a location with a large variety of sounds and interactions, but governed by well-defined social norms so as to generate data as close as possible to a potential critical working environment such as bank branches. The test set was generated by placing artificial anomalies at different random instants of the additional day, so as to have a simple and reliable labelling process. Different classes of anomalies have been recorded providing samples of both human interactions, like crying or screaming, and mechanical sounds (electronic screwdriver, drill, grinder etc). These additional sounds have been later mixed into the original audio samples exploiting the pydub library.

The network is implemented using the Keras framework [28][29]. Both the encoding and decoding LSTMs are composed by 512 units for each gate (with sigmoid activations), whereas the static AE contains 30 units in its hidden layer (with ReLU activation). A validation set (30% of training samples) has been held out to stop the optimization when the reconstruction error, evaluated as the mean squared error (MSE) between the input and the output sequences, stops to decrease. In table 2 we report the Area under the ROC (AUC) obtained in each setting to evaluate the general behaviour of each model. To give an idea of the classification performances, in table 3 we reported the F1-score obtained when selecting the threshold at the point in which the ROC crosses the diagonal.

Table 2. Area Under the ROC Curve (AUC) evaluated on different running varying both the window size and the length of the sequence processed by the LSTM modules of the ANN.

	AUC			
Sequence Length s (input steps)	Window Size w (samples)			
	4096	**8192**	**16384**	**32768**
40	0.9770	0.9716	0.9537	0.9153
20	0.9755	0.9776	0.9724	0.9514
10	0.9729	0.9814	0.9810	0.9706
5	0.9721	0.9822	**0.9836**	0.9774
1	0.9480	0.9784	0.9704	0.9632

Table 3. F1-score evaluated on different running varying both the window size and the length of the sequence processed by the LSTM modules of the ANN.

	F1-score			
Sequence Length s (input steps)	Window Size w (samples)			
	4096	**8192**	**16384**	**32768**
40	0. 8680	0. 8653	0. 8194	0. 7213
20	0. 8537	0. 8734	0. 8678	0. 8115
10	0. 8449	0. 8810	0. 8902	0. 8648
5	0. 8461	0. 8808	**0. 8946**	0. 8829
1	0. 7787	0. 8684	0. 8413	0. 8264

The results are plotted in figure 2, while figure 4 depicts the classification performances varying the size of the features extraction window and the length of the network input sequence while keeping w/s constant. In figure 3 we show the behaviour of the reconstruction error from different models in presence of an anomaly. In figure 2 we can observe how in general the RNN performs better than the static AE. Best performances are achieved for each w at $s = 5$, except from the shortest window with $w = 4096$. In this case, increasing the length of the processed sequence helps to get close to the results obtained by using wider windows. On the other hand, for larger windows the extension of the input sequence leads in general to a degradation of the performances, especially for the largest one with $w = 32768$. This could be due to the harder training task implied by the longer input sequence, and we leave as a future work to analyse if this issue can be mitigated by increasing the number of units in the LSTM layers. In the right plot of figure 3, it is possible to observe how the processing of long sequence produces a smoother reconstruction error, which implies a reduction of the precision on the anomalies boundaries. Another factor to be considered is the width of the temporal window expressed by the parameter w. Indeed, the anomalies are collapsed into a smaller number of frames with a large window, raising the detection difficulty. An opposite trend can be observed in the left plot of figure 3. In the case $s = 1$ the reconstruction error shows a higher variability, proving that, as a matter of fact, the static model seems to provide a poorly representation. In figure 4 we reported a comparison among the results obtained by processing configurations with constant rate w/s (following the diagonals of table 1). The sequential approach seems to introduce fundamental improvements even in these sense, still confirming the issue of processing long input sequences.

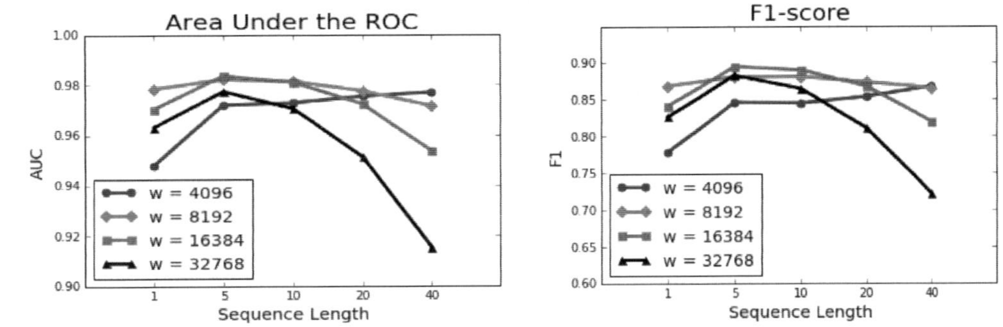

Figure 2. Area Under the ROC (AUC) and F1-score, respectively on the left plot and the right one, obtained varying the sequence length s and the window size w.

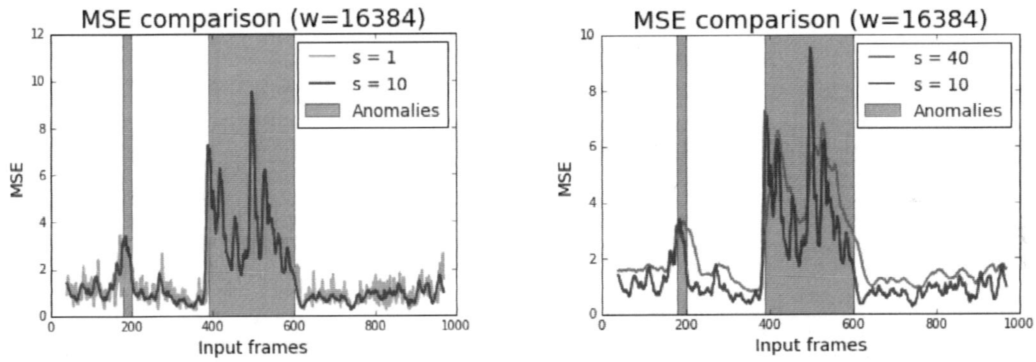

Figure 3. Reconstruction error on a slice of input stream in presence of anomalies with both short and long durations, obtained with models trained varying the input sequence length s.

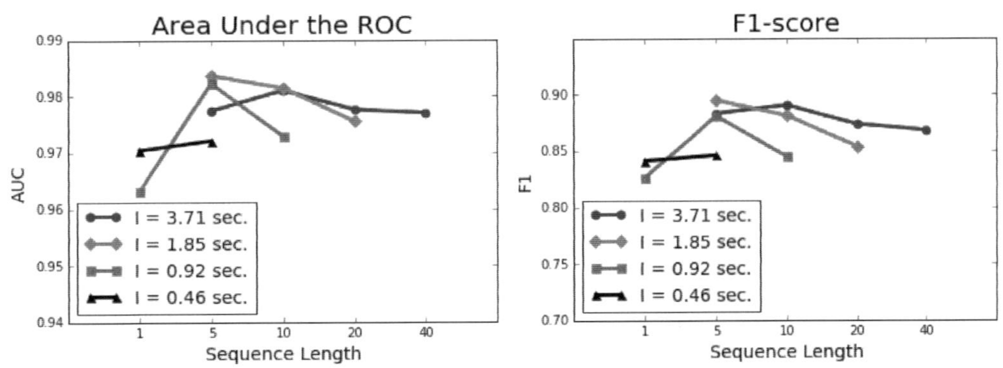

Figure 4. *Area Under the ROC* (AUC) and *F1-score,* respectively on the left plot and the right one, obtained varying the sequence length s and the window size w. The comparing is reported by keeping w/s constant.

4. Conclusions

This work studied the application of a sequence-to-sequence model to the problem of identifying abnormal events in an audio input. The proposed solution is purely unsupervised and requires only a stream of normal audio to be trained. This paper also presents an investigation about the importance of considering the temporal sequence when representing an audio signal. The showed results confirm the expected hypothesis that sequential models should work better when dealing with data characterized by an explicit temporal order. Indeed, the static approach, obtained by processing single input frames of a given temporal size, seems to provide a worse representation in each analyzed setting, even when the effectively processed interval of time is the same as the sequential approach. This result seems to be general w.r.t. the used feature representation, since the recurrent modules do not receive any additional information, except from the one produced by correlating consecutive input frames. The study suggests that there is an optimal length in the processed sequences as increasing too much the length of the input sequences leads to a degradation in the classification accuracy. As future work, we plan to investigate alternative architectures limiting the reconstruction of the output sequence, which is very hard to train. For example, once the encoding of the sequence is obtained, the decoder can be used to generate only the last input frames instead of the whole input sequence.

Acknowledgments
We wish to thank Banca Monte dei Paschi di Siena (Siena, Italy) that provided the founding and the main proposals for this research.

References

[1] Sharan R V and Moir J T 2016 Neurocomputing 200 22-34
[2] Orio N et al. 2006 Foundations and Trends in Information Retrieval 1 1-90
[3] Cowling M and Sitte R 2003 Pattern Recognition Letters 24 2895-907
[4] Chachada S and Kuo J 2013 Signal and Information Processing Association Annual Summit and Conference (APSIPA) 1-9.
[5] Crocco M, Cristani M, Trucco A, and Murino V 2016 Computing Surveys (CSUR) 48 52
[6] Choi W et al. 2012 Advanced Video and Signal-Based Surveillance (AVSS) 118-123
[7] McLoughlin I et al. 2015 IEEE/ACM Transactions on Audio, Speech, and Language Processing 23 540-52
[8] Pimentel M, Clifton D, Clifton L and Tarassenko L 2014 Signal Processing 99 215-49
[9] Williams G et al. 2002 International Conference on Data Mining 709-12
[10] Bishop C M 1994 IEE Proceedings-Vision, Image and Signal processing 141(4) 217-22
[11] Tarassenko L et al. 1995 4th International Conference on Artificial Neural Netowrks 442-7
[12] Foggia P et al. 2016 IEEE Transaction on Intelligent Transportation Systems 17(1) 279-88
[13] Ntalampiras S, Potamitis I and Fakotakis N 2011 IEEE Transaction on Multimedia 13 713-9
[14] Graves A et al. 2009 IEEE Transaction on pattern analysis and machine intelligence 855-68
[15] Sundermeyer M et al. 2012 13th Annual Conference of the International Speech Communication Association
[16] Sutskever I et al. 2014 Advances in neural information processing systems 3104-12
[17] Hochreiter S and Schmidhuber J 1997 Neural Computaion 9 1735-80
[18] Marchi E et al. 2017 Computational intelligence and neuroscience
[19] Werbos P 1990 Proceedings of the IEEE 78(10) 1550-60
[20] Hochreiter S et al. 2001 A field guide to dynamical recurrent neural netoworks. IEEE Press
[21] Bengio Y 1994 IEEE transaction on neural networks 5(2) 157-66
[22] Kalchbrenner N and Blunsom P 2013 EMNLP 3(39) 413
[23] Graves A 2013 arXiv preprint (arXiv:1308.0850)
[24] Cho K et al. 2014 arXiv preprint (arXiv:1406.1078)

[25] Ranzato M et al. 2006 Advances in Neural Information Processing Systems 19 1137-44
[26] McFee B et al. 2015 Proceedings of the 14th python in science conference 18-25
[27] Antti et al. 2006 IEEE Transaction on Audio, Speech and Language Processing 14 321-9
[28] Chollet F et al. 2015 https://github.com/fchollet/keras
[29] Pedregosa F et al. 2011 Journal of Machine Learning Research

IOP Conf. Series: Materials Science and Engineering **261** (2017) 012010 doi:10.1088/1757-899X/261/1/012010

Smooth Optimal Control for a Class of Switched Systems Based on Fuzzy Theory and PSO

Esmaeil Atashpaz-Gargari

School of Engineering and Computing, National University, San Diego, CA.

egargari@nu.edu

Abstract. This paper proposes an approach to smooth optimal control design of two-point boundary value problem of switched systems with fuzzy operating regions. The switched system is modeled via a T-S fuzzy system, and then the fuzzy inference method of T-S fuzzy system is used to transform the switched system to a nonlinear system. For this nonlinear system, an optimal controller is designed. The control signal is a fuzzy control with weighted average defuzzifier. The fuzzy sets are used to partition the time space and a Particle Swarm Optimization (PSO) algorithm is proposed to optimize the weights. Simulations results demonstrate the applicability and effectiveness of the proposed method.

1. Introduction

Since many practical dynamical systems are of switching in nature, recently the control and analysis of these systems have been the focus of studies [1]. In this paper, we focus on the state dependent switched systems. In a state dependent switched system the continuous state space is partitioned into a finite or infinite number of operating regions by means of a family of switching surfaces, or guards. Whenever the system trajectory hits a switching surface, the continuous state jumps instantaneously to a new value, specified by a mapping. While some papers have been published on the analysis of such systems [2, 3], there are few works on optimal control of these systems.

In many practical applications, the switching surfaces are not known exactly and they are determined by linguistic variables based on human knowledge. Thus, in practice, we may have systems with fuzzy operating regions.

This paper proposes a novel approach for optimal control of a class of state dependent switched systems with fuzzy operating regions. Here, the switched system is firstly expressed via Takagi-Sugeno (T-S) fuzzy rules [4], and then by following the fuzzy inference method of T-S fuzzy systems, a nonlinear system is derived from the T-S fuzzy rules. Now, we have a nonlinear optimal control two-point boundary value problem. An optimal fuzzy control with weighted average defuzzifier is used, where the fuzzy sets are selected to partition the time space and a Particle Swarm Optimization (PSO) algorithm is proposed to optimize the weights. To demonstrate the applicability of the proposed method, it is applied to a switched two-point boundary value problem. Simulation results show the effectiveness of the proposed method.

This paper is organized as follows. Section 2 states problem formulation, assumptions, and preliminaries. Transformation based on T-S fuzzy systems is presented in Section 3. A brief description of the Particle Swarm Optimization algorithm is presented in Section 4. Section 5 describes and proposes the optimal controller, while in Section 6 a simulation example is demonstrated and finally, Section 7 concludes the main advantages of the proposed method.

2. Problem Formulation and Preliminaries

Let x be a vector in n-space $\mathbb{R}^n$, u be a vector in m-space $\mathbb{R}^m$, and t be a real variable. $J = [t_a, t_b]$, with $t_a < t_b$ is a real closed interval. The interior of this interval will be denoted by J^o, $J^o = (t_a, t_b)$. This is the time interval in which the controlled system will evolve. The set A in $\mathbb{R}^n$ is bounded, closed, pathwise-connected, which the trajectory of the system is constrained to stay in it for $t \in J$. Two elements of A, x_a and x_b, are the initial and final states of the trajectory of the controlled system. U is a bounded closed subset of $\mathbb{R}^m$. This is the set in which the control function are to take values. Let g_i, $i = 1, \cdots, k$, be k continuous functions from $\Omega = J \times A \times U$ to $\mathbb{R}^n$. Consider the following state dependent switched system with uncertainty in the operating regions where they are described by human knowledge as follows:

$$\begin{aligned} \dot{x} &= g_1(t,x,u), \quad f_{11}(x) \text{ is } M_{11} \wedge \cdots \wedge f_{1p}(x) \text{ is } M_{1p} \\ \dot{x} &= g_2(t,x,u), \quad f_{21}(x) \text{ is } M_{21} \wedge \cdots \wedge f_{2p}(x) \text{ is } M_{2p} \\ &\vdots \\ \dot{x} &= g_k(t,x,u), \quad f_{k1}(x) \text{ is } M_{k1} \wedge \cdots \wedge f_{kp}(x) \text{ is } M_{kp} \end{aligned} \tag{1}$$

where, each M_{ij}, are fuzzy sets in A, k is the number of switched systems, $f_{ij}(x)$ are known nonlinear functions which are functions of states and $\wedge$ is fuzzy intersection.

Assumption 2.1
The trajectory functions $t \in J \rightarrow x(t) \in A$ are absolutely continuous and the control function $t \in J \rightarrow u(t) \in U$ is Lebesgue measurable.

Assumption 2.2
Without loss of generality it is assumed that the membership functions are normalized, thus we have $\sum_{i=1}^{k} z_i(f_i(x)) = 1$, where

$$\begin{aligned} f_i(x) &= [f_{i1}(x), f_{i2}(x), \cdots, f_{ip}(x)] \\ w_i(f_i(x)) &= \prod_{j=1}^{p} \mu_{M_{ij}}(f_{ij}(x)) \\ z_i(f_i(x)) &= \frac{w_i(f_i(x))}{\sum_{i=1}^{k} w_i(f_i(x))} \end{aligned} \tag{2}$$

and $\mu_{M_{ij}} : x \rightarrow [0,1]$ is the membership function of fuzzy set M_{ij}.

The objective is to minimize the following functional:

$$\int_J f_0(t,x,u)\,dt \tag{3}$$

such that, $f_0 : \Omega \rightarrow \mathbb{R}$ ($\Omega = J \times A \times U$) be a continuous function, the control function $u(.)$ takes values in the set U, and the boundary conditions are $x(t_a) = x_a, x(t_b) = x_b$. The cost functional (3) is written in a general form, and there is not any mathematical restrictions on the integrand function. However, in practice, the integrand is usually a positive semi-definite function, for every possible state and input combination.

3. Transformation based on T-S Fuzzy Systems

The fuzzy model proposed by Takagi-Sugeno (T-S) [5] is described by fuzzy IF-THEN rules which represent local input-output relations of a nonlinear system. The main feature of a T-S fuzzy model is to express the local dynamics of each fuzzy implication (rule) by a linear/nonlinear system model. The overall fuzzy model of the system is achieved by fuzzy "blending" of the linear/nonlinear system models. It is proved that many nonlinear dynamic systems can be represented by T-S fuzzy models (i.e., fuzzy models are universal approximators) [6]. The switched system (1) can be modeled via a T-S fuzzy logic system with the following IF-THEN rules:

Rule i, where $i = 1, 2, \cdots, k$:

$$\begin{aligned} &\text{IF } f_{i1}(x) \text{ is } M_{i1} \text{ and} \ldots \text{ and } f_{ip}(x) \text{ is } M_{ip} \\ &\text{THEN } \dot{x}(t) = g_i(t, x, u) \end{aligned} \tag{4}$$

Given the pair of $(x(t), u(t))$, the output of the fuzzy system derived from the original switched state dependent system is inferred as the following nonlinear system:

$$\dot{x}(t) = \sum_{i=1}^{k} z_i(f_i(x)) g_i(t, x, u) = h(t, x, u) \tag{5}$$

for all t, where $z_i(f_i(x))$ is defined in Assumption 2.2 and based on this assumption we have $\sum_{i=1}^{k} z_i(f_i(x)) = 1$. Thus, the main problem is converted to the following optimal control problem:

$$\begin{aligned} &\int_J f_0(t, x, u)\, dt \\ &\text{s.t.} \\ &\dot{x} = h(t, x, u) \qquad t \in J,\ x \in A,\ u \in U \\ &x(t_a) = x_a \\ &x(t_b) = x_b \end{aligned} \tag{6}$$

4. Particle Swarm Optimization

Kennedy and Eberhart [7] introduced the Particle Swarm Optimization (PSO) algorithm, which is inspired by social behaviors observed in fish schools and bird flocks. PSO is mainly based on Swarm Intelligence paradigm, and also has many common properties with Evolutionary Computation methods. In PSO, particles are individuals, representing solutions of an optimization problem, and moving in the search space, obeying some permanent rules, to find the global optimum. Since its inception, PSO has been used for solving diverse range of optimization problems in the area of engineering and science [8, 9, 10] and different variants of the algorithm are proposed [11, 12].

A particle memorizes its best experience, namely personal best, and uses its own best performance and best performance of the whole swarm of particles, known as global best, to find a better point in the search space. Particles have five basic properties: Current Position, Velocity, Current Objective Value, Best Position ever found (or Personal Best), and Objective Value corresponding to the Best Position.

Consider a minimization problem, with cost function $f : \mathbb{R}^n \rightarrow \mathbb{R}$. Hence solutions of this problem are real valued n-dimensional points. Assume that position of particle i, at iteration t, is represented by $x_i[t]$, and $x_{i,j}[t]$ stands for the value of j-th element of the position. Similarly, the velocity of particle i, at iteration t, and the value of its j-th element, are shown by $v_i[t]$ and $v_{i,j}[t]$, respectively. The symbols $p_i[t]$, $p_{i,j}[t]$, $g[t]$ and $g_j[t]$, stand for best position of particle i, j-th element of best position of particle i, global best position, and j-th element of global best position, respectively. The update rules for PSO are defined by:

$$v_{i,j}[t+1] = w v_{i,j}[t] + c_1 r_1 (p_{i,j}[t] - x_{i,j}[t]) + c_2 r_2 (g_j[t] - x_{i,j}[t]) \tag{7}$$

and

$$x_{i,j}[t+1] = x_{i,j}[t] + v_{i,j}[t+1] \tag{8}$$

where w is the inertia factor, c_1 is personal learning factor, c_2 is global learning factor, and r_1 and r_2 are random numbers uniformly distributed in the range [0,1].

In this paper, the values of inertia factor and learning coefficients are chosen according to [13], as follows:

$$\begin{aligned} w &= \frac{2}{\phi - 2 + \sqrt{\phi^2 - 4\phi}} \\ c_1 &= \phi_1 w \\ c_2 &= \phi_2 w \end{aligned} \tag{9}$$

where ϕ_1 and ϕ_2 are positive real constants, and $\phi = \phi_1 + \phi_2$. A good choice for ϕ_1 and ϕ_2, is $\phi_1 = \phi_2 = 2.05$. Therefore according to (8), the values of PSO parameters are set as follows: $w \approx 0.7298$, and $c_1 = c_2 \approx 1.4962$. These values are used in the simulations of this paper.

5. Proposed Method

In this paper the optimal controller is the output of a Mamdani type fuzzy system. The fuzzy system contains fuzzifier, fuzzy rule base, inference engine and the defuzzifier. The input of the fuzzy system is selected as and the fuzzy rule base is built by the following fuzzy IF-THEN rules

$$\text{IF } t \text{ is } A_l \text{ THEN } u \text{ is } B_l, \qquad l = 1, \cdots, M \tag{10}$$

where, A_l and B_l are Gaussian fuzzy sets such that A_l, $l = 1, \cdots, M$ are used to partition the compact set J, and B_l, $l = 1, \cdots, M$ are linguistic description of control signal u. Using singleton fuzzifier, product inference engine and center average defuzzifier the output of the fuzzy system which is also the control signal is obtained as

$$u(t) = \frac{\sum_{l=1}^{M} w_l \mu_{A_l}(t)}{\sum_{l=1}^{M} \mu_{A_l}(t)} \tag{11}$$

where, μ_{A_l} is the membership function corresponding to fuzzy set A_l and w_l is the center of the membership function corresponding to fuzzy set B_l. Thus we have a smooth control signal. On the other hand, the control signal $u(t)$ based on (6) should be designed such that the cost functional

$$\int_J f_0(t, x, u)\, dt \tag{12}$$

is minimized and also the following boundary condition is satisfied:

$$x(t_b) = x_b \tag{13}$$

Therefore, the center of the output membership functions of the fuzzy system (w_l) are assumed free and tuned based on a PSO algorithm such that the following cost functional is minimized:

$$\alpha\left|\frac{x(t_b)-x_b}{x_b}\right|+\int_J f_0(t,x,u)\,dt \tag{14}$$

where, it contains (12) and also condition (13) is added to (14) as a penalty term. The penalty weight parameter α is selected as a big positive constant. Initial state of the system is also a constraint in (6) and it is considered and satisfied in the simulation and optimization process.

6. Simulation Example

Consider the following nonlinear switched optimal control problem:

$$\begin{aligned}
&\min I(u(.)) = \int_0^1 u^2(t)\,dt \\
&\dot{x} = \frac{1}{2}x^2 \sin x + u, \qquad x \text{ is Positive} \\
&\dot{x} = x^3 + u, \qquad x \text{ is Negative} \\
&x(0) = -0.5 \\
&x(1) = 0.8
\end{aligned} \tag{15}$$

where, $x \in A = [-1,1]$, $J = [0,1]$ and $u \in U = [-1,8]$, $M_1 = \text{Positive}$, $M_2 = \text{Negative}$ and

$$\begin{aligned}
\mu_{M_1}(x) &= \frac{1}{1+\exp(-\frac{x}{0.01})}, \\
\mu_{M_2}(x) &= \frac{\exp(-\frac{x}{0.01})}{1+\exp(-\frac{x}{0.01})}
\end{aligned} \tag{16}$$

Now, we use the following T-S fuzzy rules:

$$\begin{aligned}
&\text{IF } x \text{ is } M_1 \text{ THEN } \dot{x} = \frac{1}{2}x^2 \sin(x) + u \\
&\text{IF } x \text{ is } M_2 \text{ THEN } \dot{x} = x^3 + u
\end{aligned} \tag{17}$$

Now, by following the fuzzy inference method of T-S fuzzy systems the fuzzy blending of the rules in (17) becomes the following nonlinear system:

$$\begin{aligned}
\dot{x} &= \frac{1}{1+\exp(-\frac{x}{0.01})}\left(\frac{1}{2}x^2 \sin x + u\right) + \frac{\exp(-\frac{x}{0.01})}{1+\exp(-\frac{x}{0.01})}(x^3+u) \\
&= \frac{x^2}{1+\exp(-\frac{x}{0.01})}\left[\tfrac{1}{2}\sin x + x\exp(-\frac{x}{0.01})\right] + u
\end{aligned} \tag{18}$$

Therefore, we have the following optimal control problem:

IOP Conf. Series: Materials Science and Engineering **261** (2017) 012010 doi:10.1088/1757-899X/261/1/012010

$$\min I(u(.)) = \int_0^1 u^2(t)\,dt$$
$$\dot{x} = \frac{x^2}{1+\exp(-\frac{x}{0.01})}\left[\tfrac{1}{2}\sin x + x\exp(-\frac{x}{0.01})\right] + u \tag{19}$$
$$x(0) = -0.5$$
$$x(1) = 0.8$$

Based on the cost functional (14), the nonlinear optimal control (19) can be written as

$$\min I(u(.)) = 1000\left|\frac{x(1)-0.8}{0.8}\right| + \int_0^1 u^2(t)\,dt$$
$$\dot{x} = \frac{x^2}{1+\exp(-\frac{x}{0.01})}\left[\tfrac{1}{2}\sin x + x\exp(-\frac{x}{0.01})\right] + u \tag{20}$$

Using the proposed method the smooth optimal control and the corresponding optimal trajectory is depicted in Figure. 1 and 2, respectively. The smoothness of the fuzzy switching behaviour of the system itself and the smooth fuzzy control input results in the smooth response of the system, shown in Figure 1. In contrast, for normal non-fuzzy switching systems, due to discontinuous and crisp switching rules, sudden changes and jumps in the response can be observed.

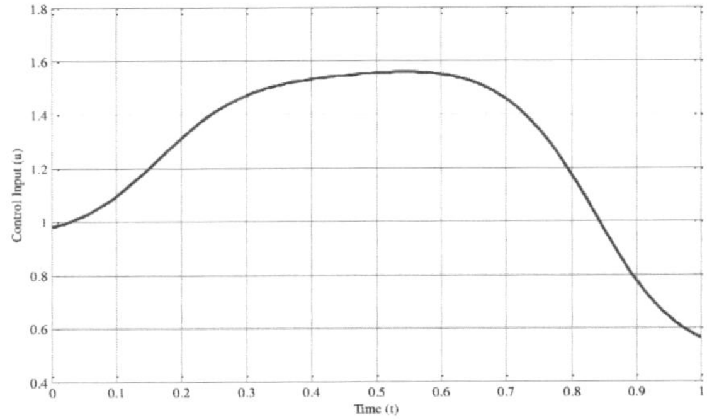

Figure 1. Smooth Optimal Control Input resulted from PSO

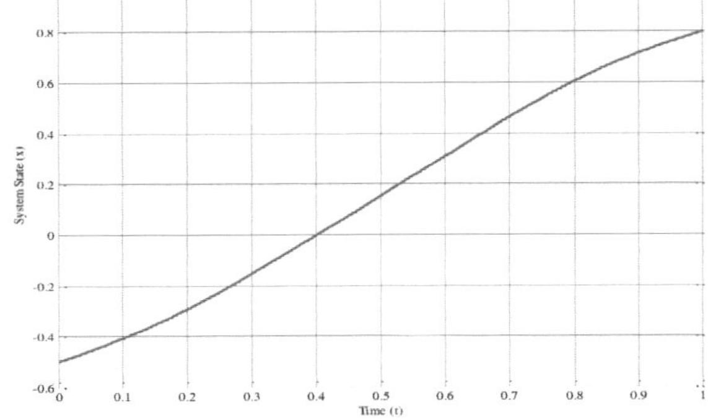

Figure 2. Optimal State Trajectory corresponding to the Smooth Optimal Control Input shown in Figure 1.

IOP Conf. Series: Materials Science and Engineering **261** (2017) 012010 doi:10.1088/1757-899X/261/1/012010

7. Conclusion

In this paper we proposed a new approach for optimal control of a class of state dependent switched systems with fuzzy based operating regions. First the switched system was modeled via T-S fuzzy systems, and then by following the fuzzy inference method of T-S fuzzy system, the switched system was transformed to a nonlinear system. In the next step, a smooth control built by a fuzzy system with weighted average defuzzifier was proposed. To optimize the weights such that a desired cost functional is minimized a PSO algorithm was used. To show the effectiveness of the proposed method a simulation example was illustrated. In future works, we intend to propose a closed-form for the optimal control signal while the system is under modelling errors and unknown external disturbances.

References

[1] Liberzon D, 2012. Switching in systems and control. Springer Science & Business Media.
[2] Wang Y, Roohi N, Dullerud G E and Viswanathan M, 2017. Stability analysis of switched linear systems defined by regular languages. IEEE Transactions on Automatic Control, 62(5), pp.2568-2575.
[3] Lin X, Li S and Zou Y, 2014. Finite-time stability of switched linear systems with subsystems which are not finite-time stable. IET Control Theory & Applications, 8(12), pp.1137-1146.
[4] Tanaka K, Wang HO, 2001. Fuzzy Control Systems Design and Analysis. John Wiley & Sons Inc.
[5] Takagi T, and Sugeno M, 1985. Fuzzy identification of systems and its application to modeling and control, in IEEE Transactions on Systems, Man and Cybernetics, vol. 15, pp. 116-132.
[6] Castro JL, 1995. Fuzzy logic controllers are universal approximators. IEEE transactions on systems, man, and cybernetics, 25(4), pp.629-635.
[7] Kenndy J and Eberhart RC, 1995, December. Particle swarm optimization. In Proceedings of IEEE international conference on neural networks (Vol. 4, pp. 1942-1948). IEEE Press.
[8] Bana, S and Saini, RP, 2017. Identification of unknown parameters of a single diode photovoltaic model using particle swarm optimization with binary constraints. Renewable Energy, 101, pp.1299-1310.
[9] Naderi, E, Narimani, H, Fathi, M and Narimani, MR, 2017. A novel fuzzy adaptive configuration of particle swarm optimization to solve large-scale optimal reactive power dispatch. Applied Soft Computing, 53, pp.441-456.
[10] Mousavi, SM, Bahreininejad, A, Musa, SN and Yusof, F, 2017. A modified particle swarm optimization for solving the integrated location and inventory control problems in a two-echelon supply chain network. Journal of intelligent manufacturing, 28(1), pp.191-206.
[11] Suresh, S and Lal, S, 2017. Multilevel thresholding based on Chaotic Darwinian Particle Swarm Optimization for segmentation of satellite images. Applied Soft Computing, 55, pp.503-522.
[12] Du, WB, Ying, W, Yan, G, Zhu, YB and Cao, XB, 2017. Heterogeneous strategy particle swarm optimization. IEEE Transactions on Circuits and Systems II: Express Briefs, 64(4), pp.467-471.
[13] Clerc M and Kennedy J, 2002. The particle swarm-explosion, stability, and convergence in a multidimensional complex space. IEEE transactions on Evolutionary Computation, 6(1), pp.58-73.

IOP Conf. Series: Materials Science and Engineering **261** (2017) 012011 doi:10.1088/1757-899X/261/1/012011

A Spectrum Sensing Scheme Based on Subspace Filtering

Junsheng Mu[1], Wei Yang[1, *], Xiaojun Jing[1] and Hai Huang[1]

[1]School of Information and Communication Engineering, Beijing University of Posts and Telecommunications, Beijing, 100876, China

Corresponding author: *Wei Yang (yangweibupt@bupt.edu.cn)

Abstract. Spectrum sensing (SS) has attracted much concern of researchers due to its significant contribution on the spectral efficiency. Energy Detection (ED) has been a critical method for Spectrum Sensing in Cognitive Radio Networks (CRNS) due to its low complexity and simple implement. However, noise uncertainty in ED greatly degrades the detection performance, especially under a low Signal-to-Noise Ratio (SNR). To remove noise uncertainty as much as possible, a scheme based on subspace decomposition is proposed for SS, where the received signal is decomposed into two parts: noise subspace and signal-plus-noise subspace. Then the closed-form solution of the detection and false alarm probabilities is given on the basis of the signal-plus-noise subspace in Rayleigh fading channel. The energy of the remainders after removal of noise subspace and noise contribution in signal-plus-noise subspace is used to decide whether the primary user (PU) exists by a comparison with a redesigned threshold. Eventually, some simulations based on MATLAB platform is made to validate the proposed method.

1. Introduction

Cognitive Radio (CR) [1] has been widely accepted as a vital way to relieve the limited spectrum resource and inefficient efficiency of spectrum utilization. One core technology behind a CR is Spectrum sensing (SS) [2]. SS works to acquire the spectrum usage by signal detection and processing in wireless propagation channel.

The performance of sensing spectrum holes is evaluated by the detection probability and the probability of false alarm [3]. A miss in the detection will cause the interference with PU and a false alarm will reduce the spectral efficiency [4]. The optimal detection performance occurs in the case that the detection probability is maximized subject to the constraint of the false alarm probability. According to the standard of wireless regional area networks (WRAN) [5] based on CR of 802.22 working group [6], the detection probability of propagable CR is at least for 90% while the probability of false alarm is below 10%. Practically, to the best of our knowledge, this is a hard work in the state of art considering the computational and implementation complexity.

In the literature of SS, a number of algorithms have been proposed to identify the presence of the primary signal as well as to improve the detection efficiency by inhibiting the disgusting interference in the sensing environment. Some examples of the existing proposals include Energy Detector (ED) [7], Matched Filtering (MF) [8], Covariance-Based Detection (CBD) [9] and Cyclostationary Feature Detection (CFD) [10]. MF works based on the prior knowledge of the signal and needs well time synchronization. Although the CBD based method outperforms than the others on detection performance due to its nearly independence of noise uncertainty, the required sampling frequency is much higher than the normal conditions, leading to a higher com-plexity of implementation.

Energy Detector (ED) is a classical blind spectrum sensing without the prior knowledge of signal, time synchronization and a higher sampling frequency. Moreover, ED is an optimal blind detector. However, its main drawback consists in the noise uncertainty incurred by the estimation of noise variance, especially under a lower signal-noise-ratio (SNR), where an accuracy estimation of noise variance is difficult. For a possible removal of noise uncertainty in ED, a good number of corresponding schemes are proposed, such as cooperative spectrum sensing, etc. Cooperative spectrum sensing (CSS) [11] lowers the volatility of noise variance and considers fading and multipath effect in wireless propagation channel.

To the best knowledge of the author, most of ED based methods are devoted to eliminating noise uncertainty in an indirect way and the suppression effect of noise uncertainty is not excellent enough. As a result, a spectrum sensing scheme is proposed in this paper on the basis of subspace filtering (SSF), where the received signal at the receiver passes a subspace filter [12]. The subspace filter firstly decomposes the received signal into two orthogonal subspaces, noise subspace and signal-plus-noise subspace. Noise subspace is filled with noise and sign-plus-noise subspace contains the whole signal and segmental noise. Then the noise subspace is moved away together with the noise contribution in the signal-plus-noise subspace. After subspace filtering, background noise is filtered greatly and the energy of the remainders after removal of noise subspace and noise contribution in signal-plus-noise subspace is used to decide whether the primary user (PU) exists by a comparison with a redesigned threshold. It is voted that the sensing performance will have a distinct improvement due to the filtering of environmental noise. As a supplement, simulations based on MATLAB platform verify the proposed method.

The rest of this paper is organized as follows. Section 2 presents the primary principle of subspace filtering. The main contribution of this paper is shown in Section 3, which consists in a spectrum sensing scheme based on subspace filtering. Simulation experiments and result analysis are accomplished in Section 4. Finally, Section 5 concludes this paper.

2. Subspace filtering

2.1. Model and motivation

Noise in CR communication systems is an aggregation of various independent sources including receiver device noise and environment noise [13]. Receiver device noise includes: (1) non-linearity components; (2) non-uniform and time-varying thermal noise. Environment noise includes: nearby unintended emissions and weak signals from other transmitters far away. By appealing to the Central Limit Theorem (CLT) [14], one usually assumes that the noise at the receiver is a Gaussian random variable.

Let us denote the clean primary signal at the receiver after a bass-pass filter as $s(t)$ and assume $s(t)$ is a random process of independence and identical distribution with bandwidth B, central frequency f_0 and variance σ_s^2. As discussed above, the background noise $u(t)$ is usually modeled as Gaussian, independent and identically distributed random process and $u(t)$ is independent of $s(t)$. In addition, spectrum sensing is an issue of binary hypothesis [15], as a result, the received signal at the receiver could be formulated as

$$H_1: r(t) = s(t) + u(t) \tag{1}$$

$$H_0: r(t) = u(t) \tag{2}$$

where H_1 denotes the case that the primary signal $s(t)$ exists and H_0 indicates the radio band of interest is unoccupied.

After a sampler with sampling frequency f_s at the receiver, the observed signal of two scenarios could be rewritten as

$$H_1: r(n) = s(n) + u(n) \tag{3}$$

$$H_0 : r(n) = u(n) \tag{4}$$

2.2. Subspace filterin

Subspace filtering is an active noise reduction method on speech enhancement and other aspects of signal processing due to the optimality in theory. The essence of subspace method consists in the construction of a linear estimator H, after which the received signal $r(n)$ is decomposed into two orthogonal subspaces: signal-plus-noise subspace $Hy(n)$ and noise subspace $(I - H)r(n)$. Noise subspace only contains noise and signal-plus-noise subspace contains segmental noise and the whole signal. After removal of noise subspace and noise contribution in the signal-plus-noise, the remainders $\hat{y}(n)$ could be formulated as

$$\hat{y}(n) = Hy(n) \tag{5}$$

Consequently, the main work for an optimal subspace based filtering consists in how to acquire an optimal linear estimator H_{opt}. According to the corresponding research on subspace filtering, the optimal linear estimator H_{opt} could be obtained by minimizing the signal distortion and keeping the noise residual noise under a preset threshold Δ_{th},

$$\begin{cases} \min \overline{\varepsilon_s^2} \\ \overline{\varepsilon_u^2} < \Delta_{th} \end{cases} \tag{6}$$

where R_x and R_u are the covariance matrix of x(n) and u(n), respectively. Δ_{th} is always some a number below threshold of audibility.

Due to the fact that the background noise is white, the covariance matrix of the received signal R_y could be denoted as a sum of the signal covariance matrix R_x and noise covariance matrix R_u,

$$R_r = R_s + R_u \tag{7}$$

and they share the same eigenvector matrix V. Taking eigendecomposition of matrix R_y, we shall have

$$R_r = V\Lambda_r V^T \tag{8}$$

where $\Lambda_r = diag[\lambda_1, \ldots \lambda_N]$ contains all the eigenvalues of matrix R_r. If the eigenvalues are arranged in the descending order, i.e., $\lambda_1 \geq \lambda_2 \geq \ldots \geq \lambda_N$, then we have $\lambda_k > \sigma^2$ for $k = 1, 2, \ldots, K$ and $\lambda_{K+1} = \ldots = \lambda_N = \sigma^2$. The part $\lambda_k > \sigma^2$ represents the signal subspace and the remainders are the noise subspace. Consequently, the signal covariance matrix R_s and noise covariance matrix R_u in Eq.7 could be respectively estimated as

$$R_s = V\Lambda_s V^T = V \begin{bmatrix} \lambda_1 & & & \\ & \text{O} & & \\ & & \lambda_K & \\ & & & 0 \end{bmatrix} V^T \tag{9}$$

$$R_u = V\Lambda_u V^T = V \begin{bmatrix} 0 & & & \\ & \lambda_{K+1} & & \\ & & \text{O} & \\ & & & \lambda_N \end{bmatrix} V^T \tag{10}$$

Then an approximatively equivalent expression of H_{opt} could be shown as

$$H_{opt} = V\Lambda_s(\Lambda_s + \mu\Lambda_u)^{-1}V^{-1} \tag{11}$$

In practice, the covariance matrix above-mentioned is impossible to be acquired. As a result, it is replaced with the sample covariance matrix, just as follows

$$R_r \approx \frac{1}{N-1}\sum_{i=1}^{N} r(i)r(i)^T \tag{12}$$

In theory, after removal of background noise by subspace based filtering, the remainders could be expressed as

$$\begin{cases} \hat{r}(n) = H_{opt} r(n) \approx s(n), H_1 \\ \hat{r}(n) \approx 0, H_0 \end{cases} \tag{13}$$

3. Spectrum sensing based on subspace filtering

In this part, the proposed spectrum sensing scheme based on subspace filtering is mainly discussed. Let us define the energy of $s(n)$ as

$$E_s = E(|s(n)|^2) \approx E(|H_{opt} s(n)|^2) \tag{14}$$

Then the detection probability and false alarm probability could be formulated as

$$P_d = P(E(|H_{opt} r(n)|^2) > 0 \mid H_1) \tag{15}$$

$$P_f = P(E(|H_{opt} r(n)|^2) > 0 \mid H_0) \tag{16}$$

Eq.15-16 works well in ideal conditions when the background noise is completely removed after subspace based filtering. However, it is noted that the residual signal after subspace filtering still exists some environmental noise. Obviously, much error leaves behind to estimate the detection and false alarm probabilities within Eq.15-16.

Let us also assume the residual noise after subspace filtering is Gaussian, independent and identically distributed due to CLT. The distinction of noise distribution between the noise before subspace filtering comes with the variance and average value of Gaussian distribution. Then the revised detection probability could be expressed as

$$\begin{aligned} P_d(\varepsilon,\tau) &= P\left(\frac{1}{N}\sum_{n=1}^{N} |y(n)|^2 > \varepsilon \middle| H_1\right) \\ &= Q\left(\left(\frac{\varepsilon}{\sigma_u^2} - 1 - \gamma\right)\sqrt{\frac{\tau f_s}{2\gamma+1}}\right) \end{aligned} \tag{17}$$

$$Q(x) = \frac{1}{\sqrt{2\pi}}\int_x^{\infty} \exp(-t^2/2) \tag{18}$$

where τ is the sensing time, f_s is the sampling frequency of SU, $N = \tau f_s$ is the sampling number of SU, $\gamma = \sigma_s^2 / \sigma_u^2$ (σ_u^2 is the variance of noise after subspace filtering)and ε is the predetermined energy threshold.

The corresponding false alarm probability could be described as

$$\begin{aligned} P_f(\varepsilon,\tau) &= P\left(\frac{1}{N}\sum_{n=1}^{N} |u_1(n)|^2 > \varepsilon \mid H_0\right) \\ &= Q((\varepsilon - \mu_0)/\sqrt{\sigma_0^2}) \end{aligned} \tag{19}$$

where $\mu_0 = \sigma_u^2$,$\sigma_0^2 = \sigma_u^4 / N$ are the expectation and variance of the original ED method for SS, and$u_1(n)$ is the residual noise after subspace filtering.

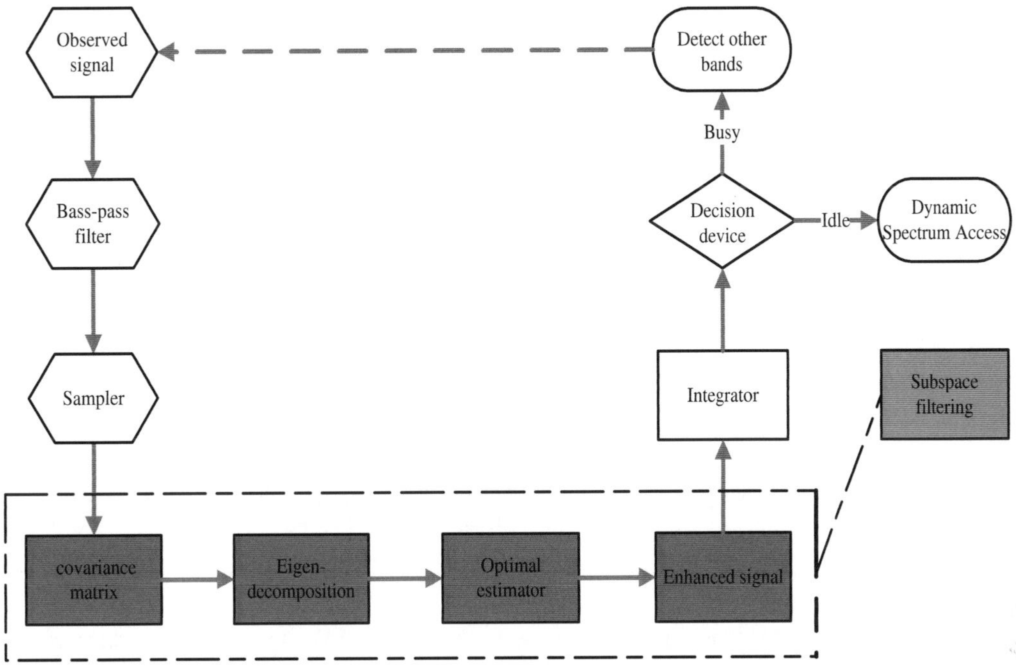

Figure.1. the flowchart of subspace based spectrum sensing

As Figure.1 shown, the general procedures for subspace based spectrum sensing are listed. The green pane denotes the subspace filtering, which corresponds to Eq.5-11. The decision device works dependent on Eq.15-16 or Eq.17-19. Note that when decision device considers current radio band to be busy, a CR will start to monitor and sense other bands for opportunistic occupation within a certain amount of time.

4. Simulation and analysis

In this section, Simulations are made to check the proposed method. The simulation band is conducted at Very high frequency (VHF), where the carrier frequency is set as the sampling frequency of SU is $1.0 \times 10^8 hz$; the sampling frequency f_s of SU is $3.0 \times 10^8 hz$ and the corresponding sampling number N is 100. Considering the sensing time and detection performance simultaneously, $N = 100$. In the simulation, the OFDM signal is firstly generated and then white Gaussian noise with mean zero and variance one is added into the OFDM signal. After the sampling and energy normalization of the noise-only case and the signal-plus-noise case, subspace based filtering is applied to them, respectively. Ultimately, we calculate the detection and false alarm probability by the corresponding energy comparisons with the presupposed threshold.

In Figure.2, a comparison is made between SSF and the other spectrum sensing schemes. Note that SSF (1) represents the sensing scheme in Eq.15-16, where the remainder energy after subspace filtering is compared with 0 to determine whether PU exists. SSF (2) denotes the spectrum hole sensing with Eq.17-19, where the residual noise after subspace filtering is assume to be Gaussian distribution. In addition, 3-MV describes a cooperative sensing scheme named Majority Voting (MV) and 3 indicate the number of secondary user (SU). From Figure.2, SSF (1) and SSF (2) have a distinct advantage over ED. Compared with 3-MV, SSF (1) has an approximate performance and SSF (2) is superior to it. The superiority of SSF on performance derives from less noise in the signal after subspace filtering. Consequently, the proposed method is competent in monitoring and sensing spectrum holes.

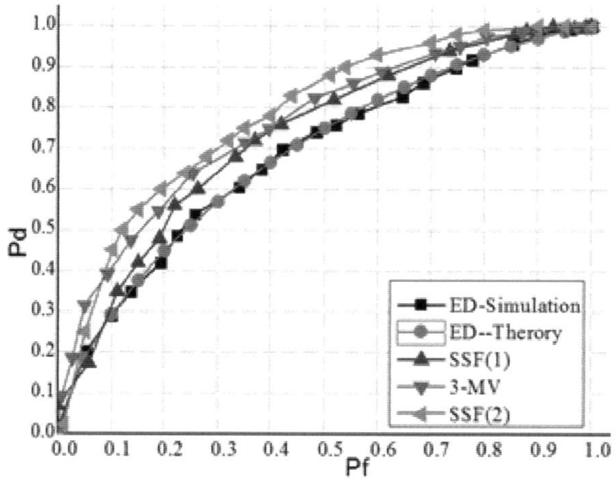

Figure 2. ROC curve comparison in AWGN channel (SNR=10dB).

As a supplement, Figure.3 gives the detection probability for a given false alarm probability ($P_f = 0.1$) when $SNR = -15dB$. From the histogram, the performance of SSF still outperforms ED and SSF (2) has the best performance in contrast with ED and SSF (1), which corresponds to the simulation made in Figure.2.

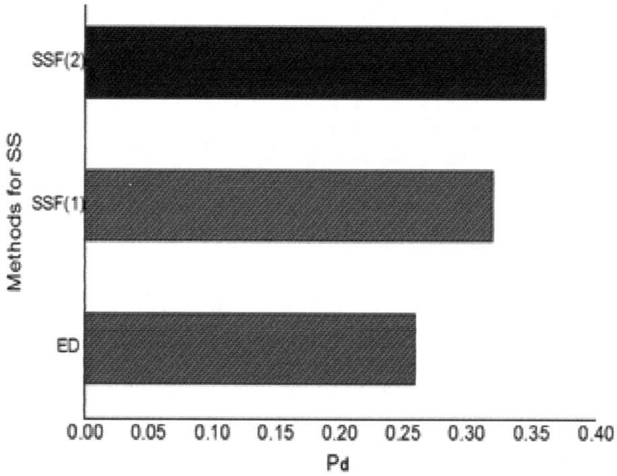

Figure 3. The detection probability for a given false alarm probability.

To sum up, the simulation results in Figure.2-3 fits our prediction and the theoretical derivation in Section 3. After subspace based filtering, the decision has less influence from environmental noise. Hence, the decision is more accurate than before.

5. Conclusions

In this paper, a subspace filtering based spectrum sensing scheme is proposed to monitor and sense spectrum holes as accurate as possible. The proposed method starts with subspace filtering for the received signal at the receiver. Then the energy of remainder signal is considered as the criterion to determine whether the radio frequency band is idle. The simulation based on MATLAB platform has been verified the proposed method.

References

[1] Iii J M, Jr G Q M. Maguire, G.Q.: Cognitive radio: making software radios more personal. IEEE Pers. Commun. 6(4), 13-18[J]. 1999, 6(4):13-18.
[2] Haykin. Cognitive radio: brain-empowered wireless communications. IEEE Journal on Selected Areas in Communications, 2005, 23(2):201-220.
[3] Naeem M, Khwaja A S, Anpalagan A, et al. Green Cooperative Cognitive Radio: A Multiobjective Optimization Paradigm[J]. IEEE Systems Journal, 2017, 10(1):240-250.
[4] Ren J, Zhang Y, Zhang N, et al. Dynamic Channel Access to Improve Energy Efficiency in Cognitive Radio Sensor Networks[J]. IEEE Transactions on Wireless Communications, 2016, 15(5):3143-3156.
[5] Xiong W, Mukherjee A, Kwon H. MIMO Cognitive Radio User Selection with and without Primary Channel State Information[J]. IEEE Transactions on Vehicular Technology, 2016, 65(2):1-1.
[6] Zhang D, Chen Z, Ren J, et al. Energy Harvesting-Aided Spectrum Sensing and Data Transmission in Heterogeneous Cognitive Radio Sensor Network[J]. IEEE Transactions on Vehicular Technology, 2017, PP(99):1-1.
[7] Liu X, Jia M, Tan X. Threshold optimization of cooperative spectrum sensing in cognitive radio networks [J]. Radio Science, 2016, 48(1):23-32.
[8] Liao Y, Song L, Han Z, et al. Full duplex cognitive radio: a new design paradigm for enhancing spectrum usage[J]. IEEE Communications Magazine, 2015, 53(5):138-145.
[9] Mesodiakaki A, Adelantado F, Alonso L, et al. Performance Analysis of a Cognitive Radio Contention-Aware Channel Selection Algorithm[J]. Vehicular Technology IEEE Transactions on, 2016, 64(5):1958-1972.
[10] Bkassiny M, Jayaweera S K, Li Y, et al. Blind Cyclostationary Feature Detection Based Spectrum Sensing for Autonomous Self-Learning Cognitive Radios[C]// IEEE International Conference on Communications Communications. IEEE, 2012:1507 - 1511.
[11] Sun H, Nallanathan A, Cui S, et al. Cooperative Wideband Spectrum Sensing Over Fading Channels[J]. IEEE Transactions on Vehicular Technology, 2016, 65(3):1382-1394.
[12] Sun C, Mu J. An eigenvalue filtering based subspace approach for speech enhancement[J]. Noise Control Engineering Journal, 2015, 63(1):36-48(13).
[13] Zeng Y, Liang Y C. Spectrum-Sensing Algorithms for Cognitive Radio Based on Statistical Covariances[J]. IEEE Transactions on Vehicular Technology, 2009, 58(4):1804-1815.
[14] Díaz-Espinosa O, Rafael D L L. Renormalization and Central Limit Theorem for Critical Dynamical Systems with Weak External Noise[J]. Journal of Modern Dynamics, 2017, 1(3):477-543.
[15] Communication S &. Cognitive Radio, Software Defined Radio, and Adaptive Wireless Systems[M]. Springer Netherlands, 2007.

IOP Conf. Series: Materials Science and Engineering **261** (2017) 012012 doi:10.1088/1757-899X/261/1/012012

Fully Convolutional Architecture for Low-Dose CT Image Noise Reduction

S Badretale[1], F Shaker[2], P Babyn[3] and J Alirezaie[4]

[1,4] Department of Electrical and Computer Engineering, Ryerson University, Toronto, ON, M5B 2K3 Canada
[2] Department of AI, Faculty of Computer Engineering, University of Isfahan, Isfahan, 81746, Iran
[3] Department of Medical Imaging, University of Saskatoon, Saskatoon, SK, S7N 0W8, Canada

[1]sbadretale@ryerson.ca, [2]f.shaker@eng.ui.ac.ir, [3]paul.babyn@saskatoonhealthregion.ca, [4]javad@ryerson.ca

Abstract. One of the critical topics in medical low-dose Computed Tomography (CT) imaging is how best to maintain image quality. As the quality of images decreases with lowering the X-ray radiation dose, improving image quality is extremely important and challenging. We have proposed a novel approach to denoise low-dose CT images. Our algorithm learns directly from an end-to-end mapping from the low-dose Computed Tomography images for denoising the normal-dose CT images. Our method is based on a deep convolutional neural network with rectified linear units. By learning various low-level to high-level features from a low-dose image the proposed algorithm is capable of creating a high-quality denoised image. We demonstrate the superiority of our technique by comparing the results with two other state-of-the-art methods in terms of the peak signal to noise ratio, root mean square error, and a structural similarity index.

1. Introduction

Computed Tomography (CT) imaging is widely used in diagnosing various medical conditions. However, the amount of x-ray needed to produce a high-quality CT image is relatively high which may result in an increasing risk of developing cancer. Therefore, it is desirable to decrease the X-ray dose for CT imaging. However, low dose CT images are highly noisy which makes them unreliable for diagnostic purposes. There have been many studies aiming to remove the noise from the low dose CT images without affecting the details of underlying image structures. X-ray CT images are reconstructed using special algorithms from the X-ray projection data. The main source of noise in X-ray projection data is the statistical fluctuations of X-ray quanta reaching the detectors. This noise follows the Poisson distribution. However, after the application of reconstruction algorithms and image processing steps to the projection data, the distribution of the noise in the resulting CT image is unknown and therefore is hard to model.

There are several approaches for denoising CT images; one is to use the original sinogram data in which the noise distribution is known to be Poisson, the other approach is to remove the noise from CT images in image domain after their reconstruction in which the noise distribution is unknown. In this research, we focus on the second approach since accessing the raw projection data is not always possible.

There has been a number of works regarding the denoising of CT images in the image domain. Ghadrdan et al. used dictionary learning combined with wavelet transform [1]. They used wavelets to extract the most suitable features in images to obtain accurate dictionary atoms for denoising algorithm. To improve the computational complexity, the number of clusters was decreased. They compared their method with the clustering based sparse representation (CSR) and the K-SVD method. Their method shows some improvement in computational complexity and noise reduction over the compared methods. Denoising of CT images using dictionary learning was also studied by Khodabandeh et al. [2].

In another study by Hashemi et al. [3], a noise confidence region evaluation (NCRE) method was used in combination with the block matching and 3Dfiltering (BM3D) method to adaptively estimate the noise properties instead of using an independent identical additive Gaussian noise model. The NCRE adaptively updates the regularization parameters at the end of each iteration by validating the result of that iteration.

Deep learning is an emerging area in AI with many successful applications in recent years. Unlike traditional multilayer NNs, which are fully connected, deep neural networks have sparse connections. Convolutional neural networks (CNN) are a subset of deep neural networks. The problem of denoising can be considered as a designing task in which an appropriate filter is constructed to remove the noise. This view aligns well with CNN in which a combination of filters in different layers is learned through real examples. The nonlinear property of the overall filter lends itself to complex distributions, which otherwise are hard to design.

Using the methods that rely on learning will be more appropriate in this scenario since the noise model is learned from the actual data rather than by estimating the complicated noise model. In this study, deep learning is used in denoising of the CT images.

In section 2, the details of our algorithm are presented. Section 3 contains the results and comparisons of our techniques and finally, in section 4 some concluding remarks are provided.

2. Materials and Methods

2.1. Problem Formulation

In low-dose CT imaging, the noise makes the edges and fine image features blurred and distorted. The goal of low-dose CT image denoising is to map the noisy and distorted image onto the clean image such that the noise is reduced or fully removed. In our research, we have considered actual low-dose and the corresponding normal-dose images, and the interfering normal-dose CT images by Gaussian or Poisson noise are not considered, so the noise is not modeled in this work.

Let X be the desired normal-dose CT image and Y be the low-dose counterpart. The Denoising function is defined as F.

$$F = argmin_F \|F(Y) - X\|_2^2 \qquad (1)$$

The objective here is to determine the function that creates the closest image to X from noisy image Y. Function F can be considered as a nonlinear filter whose coefficients are not known but can be learned through training images. In this study, convolutional neural networks are used to learn such a filter.

2.2. Overall Framework

The denoising convolutional neural network is a modified CNN whose its parameters, e.g. weights and biases should be learned and calculated with the backpropagation training. An overview of the framework is shown in Figure 1.

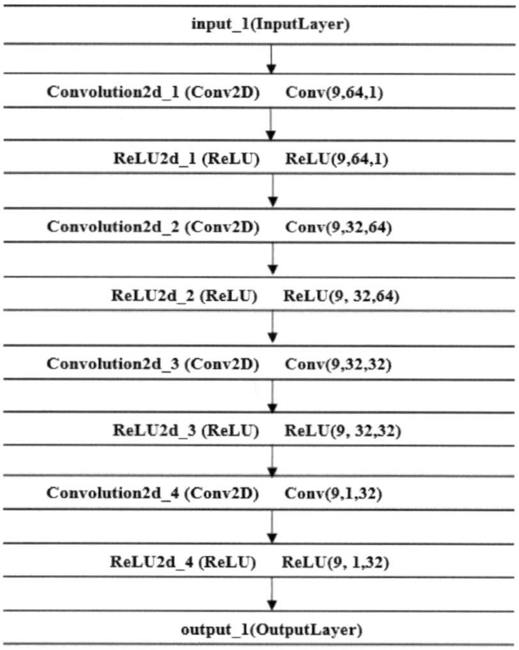

Figure 1. A network structure of the end-to-end nonlinear mapping denoising convolutional neural network.

The network that we have utilized in this study consists of four layers. Each of these layers consists of a number of convolution filters followed by an activation layer. The loss function is calculated at the end of every feed-forward step to update the parameters (weights and biases). The number of feature maps that are used from the first to the last layer is $64, 32, 32$, and 1 respectively. The size of all convolutional filters is 9×9. The input of the network is the low-dose noisy image (Y), that passes through some convolutional layers to extract sets of feature maps to produce a denoised image $F(Y)$. Hence, a number of feature maps, which require going through the activation function, form the first convolutional layer. The subsequent convolutional layers receive the output of the previous layer then pass its results through activation functions for nonlinear processing to make higher-level feature maps in order to create a corresponding noise free image. As a result, the network trained hierarchically structured feature maps from low-level (blobs, edges, etc.) to high-level (more complex and detailed shapes). Moreover, we came up with a pre-train convolutional network, which can be used as a transfer learning.

2.3. ReLU Nonlinearity

Nonlinear processing as an activation function accompanied by the pooling layer makes crucial improvements to feature extraction and makes the network rotation, shift, and scale invariant. Apart from the traditional logistic sigmoid function as an activation function, a number of nonlinear rectifying functions were introduced including the absolute positive function. Among all the activation functions, we decided to choose Rectified Linear Unit (ReLU), since several studies on deep convolutional neural networks have indicated discernible development by applying rectified nonlinear processing, and non-saturating nonlinear functions are greatly accelerated compared to saturating activation functions. By definition, the Rectified Linear Unit (ReLU) model for every input x is

$$G(x) = Max(0, x) \tag{2}$$

Mainly due to the non-saturating and linear form of ReLU, the convergence of the stochastic gradient descent is much faster than the hyperbolic tangent (tanh) and Sigmoid functions. Therefore, the training time becomes noticeably shortened. Furthermore, it does not have any reciprocal and exponential computation. That is just a simple thresholding so its implementation is much less expensive than other available alternatives. In addition, many studies have indicated that to acquire a high convergence rate RLU could be applied as an activation function [4]. We have detailed our results in Figure 2, indicating that by applying ReLU a greater PSNR can be achieved. Besides, in order to get a low error (RMSE) of 0.02 in the validation set, it needs much less number of backpropagations, 9×10^8 for ReLU as 1.6×10^8 for tanh function. The dataset consists of 500 low-dose CT images and the corresponding normal-dose CT images were taken from a patient CT chest.

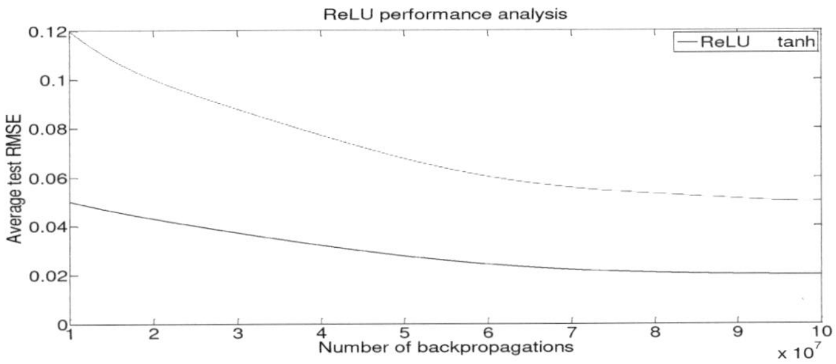

Figure 2. Comparison of tanh and ReLU as an activation function.

2.4. Implementation Details

The proposed deep convolutional neural network can be implemented by using a large set of low-dose and the corresponding normal-dose CT images. Due to a lack of sample data, low-dose CT images have been created inaccurately by adding Poisson or Gaussian noise to the normal-dose CT images. We implemented our method based on real low-dose and normal-dose datasets. We adopt widely known peak signal-to-noise ratio (PSNR), root mean squared error (RMSE), and Structural similarity (SSIM) [5] as quantitative evaluation metrics which express high correlation with the human perceptual scores to compare the result $F(Y)$ to the normal-dose image (X). Weights and biases, which are the network parameters are updated by standard stochastic gradient descent via backpropagation.

We use two different CT datasets. One dataset has 500 axial chest CT images obtained from a clinical patient. Images are from the same imaging session and it consists of two group images. One group image is taken using the higher dose of X-ray (exposure $200\ mAs$) and the other one is taken with a lower dose (exposure $25\ mAs$). Therefore, there is a one to one correspondence between images. Also, the phantom size and slice thickness is fixed. The second dataset has 500 standard phantom CT images, same number of low-dose $(25\ mAs)$ and normal-dose $(200\ mAs)$ images, containing spheres with different spacing lines with a wide variety of contrast [6]. It should be noted that creating normal-dose CT image by adding Poisson noise is not accurate. This makes the denoising algorithms impractical since the noise distribution in real low-dose CT images is not known. All the input images are in Digital Imaging and Communications in Medicine (DICOM) format.

The output images are also saved in DICOM format. The only pre-processing step is the normalization of all images between o and 1. Furthermore, we randomly cropped the training, validation and tested sets as 33×33 pixels sub-images with the stride of 14. Therefore, we obtain several thousand created small sub-images rather than overlapping patches and we do not input the original size images through the network.

The biases are initialized to zero. The filter weights of each layer for the deep convolutional neural network are initialized by drawing randomly from a Gaussian distribution with zero mean and standard deviation 0.001. Furthermore, the learning rate for the first three layers are initially set to 0.01 and is going to decrease by a factor of 0.9 every 10000 iterations. For the last convolution layer, the learning rate is initialized to 0.001 with the same decaying factor as the other layers, as well as the momentum term of 0.9.

3. Experimental Results

The proposed algorithm was validated by experiments using phantom data and real in vivo CT dataset and the results are compared with BM3D [7] and Simultaneous Sparse Coding (SSC-GSM) [8] which is claimed as the state-of-the-art image restoration by the authors.

We randomly shuffle images in each dataset in order to reflect enough diversity in the training, validation, and test data. The size of images is 512×512. The training, validation, and test data include $50\%, 20\%$, and 30% of each dataset respectively.

The proposed algorithm is implemented using the Caffe package [9] and Matlab R2016a. The processor is Intel core i7 CPU 3.4GHz and 16GB memory and the Graphics Card is GeForce GTX 1070. The training time is about 10 hours and the test time was less than a second for each image.

Table 1. indicates the average denoising results for the proposed deep convolutional neural network algorithm, BM3D, and SSC-GSM for the real CT dataset. As it is obvious in Tables 1. the proposed deep learning method provides the highest scores in all quantitative metrics. It is worth noting that the result of our proposed method is based on the output from 10^8backpropagation. Thus, it has the best performance amongst all the mentioned methods.

Table 1. Comparison of performance on the chest CT dataset.

Eval. Mat	Proposed Algorithm	SSC-GSM	BM3D
PSNR	32.112	28.754	28.298
SSIM	0.834	0.786	0.736
RMSE	0.198	0.376	0.389

Figure 3. Visualizes performance of sample output images for different methods to grasp more intuitive feeling. All the images are drawn randomly. The first two rows are from the patient CT dataset. The last row is from the phantom CT dataset. As can be observed, BM3D and SSC-GSM perform better on the phantom data due to its structured pattern of the spheres. However, our proposed method still produces better images as it removes more artifacts and creates sharper edges. That is to say, our results on the patient and phantom datasets demonstrate that clinically important details are better preserved (higher SSIM) and the noise is significantly reduced (Higher PSNR).

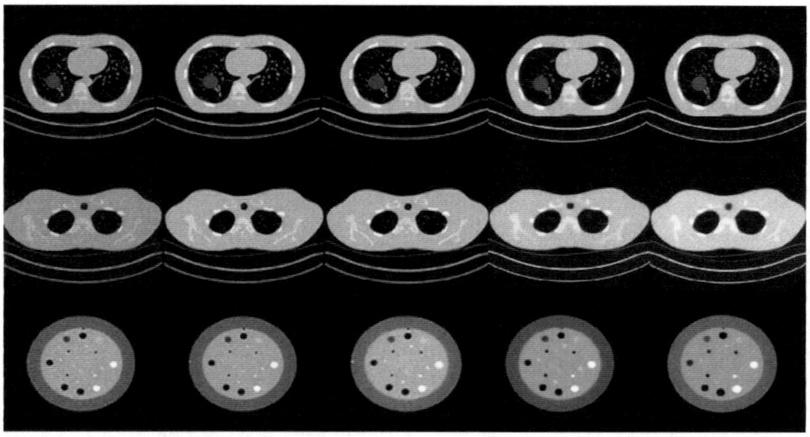

Figure 3. Visual comparison of denoising outputs. From left to right: Low-dose, Normal-dose, Proposed algorithm, BM3D method, SSC-GSM method.

In our method, the PSNR is $32.434, 34.761$, and 44.234 from top to bottom. In BM3D method, the PSNR is $29.238, 31.1$, and 41.376. And in SSC-GSM algorithm, the PSNR is $30.345, 32.964$, and 42.76. The result of the denoising methods over low-dose CT image is shown in Figure 4.

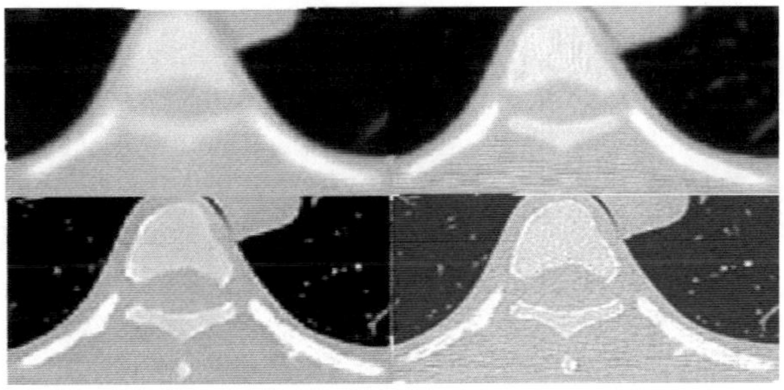

Figure 4. Zoomed images of the result of the denoising methods. Top left: BM3D, Top right: SSC-GSM, Bottom left: Proposed method, Bottom right: normal-dose.

4. Conclusion and Future work

We have introduced a novel deep learning convolutional neural network algorithm for denoising low-dose CT images. The proposed framework has a generalized formulation capable of handling a wide range of CT images with arbitrary size. We used an efficient end-to-end mapping between low-dose and normal-dose CT images. We demonstrated that the proposed algorithm has an outstanding merit due to its simplicity and surpassing of the performance of the most common methods.

We are further analyzing the effects of different strategies such as the number of convolution layers and kernel size. Also, higher quality CT images might be obtained by dataset expansion and data augmenting.

References

[1] Ghadrdan *et al.* Low-dose Computed Tomography Image Denoising based on Joint Wavelet and Sparse representation. *Engineering in Medicine and Biology Society (EMBC).* 2014. *36th Annual International Conference of the IEEE*, 3325–3328.

[2] Khodabandeh A, Alirezaie J, Babyn P, Ahmadian A. Computed Tomography Image Denoising by Learning to Separate Morphological Diversity. 2015. Telecommunications and Signal Processing (TSP). 38th International Conference on: 513–517.

[3] Hashemi S, Paul N, Beheshti S, Cobbold R. 2015. Adaptively Tuned Iterative Low Dose CT Image Denoising. 2015, 1: 1–14.

[4] Arora R, Basu A, Mianjy P, Mukherjee A, Understanding Deep Neural Networks with Rectified Linear Units. 2016. arXiv preprint arXiv:1611.01491.

[5] Wang Z, Bovik C, Sheikh H, Simmoncelli E, Image quality assessment: form error visibility to structural similarity. 2004. Image Processing, IEEE Transactions on 13, 4: 600–612.

[6] Gavrielides M, Kinnard L, Myers K, Peregoy J, Pritchard W, Zeng R, Esparza J, Karanian J, Petrick N. A resource for the assessment of lung nodule size estimation methods: database of thoracic CT scans of an anthropomorphic phantom, 2010. Optics Express, vol. 18, n.14, pp. 15244-15255.

[7] Dabov K, Foi A, Egiazarian K, Video denoising by sparse 3D transform-domain collaborative filtering. 2007. European Signal Processing Conference 16, 8, https://doi.org/10.1109/TIP.2007.901238, 145–149.

[8] Dong W, Shi G, Ma Y, Li X, Image Restoration via Simultaneous Sparse Coding: Where Structured Sparsity Meets Gaussian Scale Mixture. 2015. International Journal of Computer Vision 114, 2. https://doi.org/10.1007/s11263-015-0808-y, 217-232.

[9] Jia Y, Shelhamer E, Donahue J, Karayev S, Long J, Girshick R, Guadarrama S, Darrell T. Caffe: Convolutional Architecture for Fast Feature Embedding. 2014. ACM International Conference on Multimedia: https://doi.org/10.1145/2647868.2654889, 675–678.

IOP Conf. Series: Materials Science and Engineering **261** (2017) 012013 doi:10.1088/1757-899X/261/1/012013

Prevention 0f Unwanted Free-Declaration of Static Obstacles in Probability Occupancy Grids

Stefan Krause[1,*], **M Scholz**[2] **and R Hohmann**[1]

[1]DLR Institute of Flight Systems, Department of Unmanned Aircraft, German Aerospace Center, Germany
[2]DLR Institute of Transportation Systems, Department of Data Management and Knowledge Discovery, 38108 Braunschweig, Germany

E-mail: *stefan.krause@dlr.de

Abstract. Obstacle detection and avoidance are major research fields in unmanned aviation. Map based obstacle detection approaches often use discrete world representations such as probabilistic grid maps to fuse incremental environment data from different views or sensors to build a comprehensive representation. The integration of continuous measurements into a discrete representation can result in rounding errors which, in turn, leads to differences between the artificial model and real environment. The cause of these deviations is a low spatial resolution of the world representation comparison to the used sensor data. Differences between artificial representations which are used for path planning or obstacle avoidance and the real world can lead to unexpected behavior up to collisions with unmapped obstacles. This paper presents three approaches to the treatment of errors that can occur during the integration of continuous laser measurement in the discrete probabilistic grid. Further, the quality of the error prevention and the processing performance are compared with real sensor data.

1. Introduction

Obstacle detection for unmanned vehicles often uses approximated environment representations to fuse sensor data [1–3]. Obstacle maps are in general approximation of the real environment, to declare free or obstacle occupied spaces. A classification of various objects, e.g. trees or houses, is not the aim, the statement is rather: free spaces can be safely crossed and occupied areas are to be avoided in order to avoid collisions. Map-based obstacle detection approaches [2, 3] often use global discrete representation, which is also named occupancy grid [4]. A cell m of a discrete grid M describes a space that is free or partly occupied by an obstacle. The basic idea of an occupancy grid bases on the assumption of an ideal sensor. Each grid cell is described binary as free or occupied. A sensor measurement p is a distance measurement from a sensor origin O_S to an optically "impenetrable" surface. p describes the intersection between a ray and a surface. The cell containing p is described as occupied. The cells which are cut only by a ray, are set as free. The assumption of an ideal sensor, which can be trusted in any situation, cannot be maintained. Common approaches [1, 2, 5, 6] use an occupancy probability $p(\mathbf{p}_n)$ per cell, which is based on the deficiencies of the used sensor. This implementation is referred to as the occupancy probability grid.

The conversion of continuous measurements in a discrete grid M can generate contradictory statements about the occupancy of a cell m. The cause of contradictory statements is a rough grid resolution with respect the sensor resolution. m has a predefined edge length and it cannot be guaranteed that an obstacle will fill completely m. It is more probable that an obstacle occupies only a part of m and few rays end in m and another number of rays intersect m. Opposing occupancy probabilities $p(\mathbf{p}_n)$ for m

are the result. However, combinations of opposing $p(\mathbf{p}_n)$ are desirable at an occupancy grid to treat dynamic objects or sensor noise. Static obstacles are also influenced by the comparison of occupancy probability. Theconsequence is, based on the grid, angle and distance resolution as well as the trace of the measurements through the grid, areas in the grid are declared as free against the situation in the real world. Ignoring existing obstacles in M can lead to collisions between unmanned vehicles and obstacles. The hazard of ignoring static objects during the iterative conversion of continuous distance measurements in a discrete grid is followed named conditional unwanted free-declaration (CUFD).

The following section describes CUFD theoretically and by using real flight test data. Afterward, three approaches to covering CUFD are presented. Subsequently, a validation and a comparison of the methods show the quality of the prevention and their processing performance.

2. Problem with CUFD

A rudimentary 2D example visualizes CUFD in Figure 1. The inertial situation visualizes Figure 1a, three measurements p0 − p2 at a surface (blue) send from static sensor pose OS. The measurements are iteratively added to M. The resolution of M is dm. Figure 1b shows a subsequently incremental adding of measurements p3 up to pn and an update of M. Green cells are free spaces with an occupancy probability p(m) < 0.5. Red areas are occupied with p(m) > 0.5. White areas are unknown with p(m)= 0.5. Different intensity of colorization of freely declared cells represents different significance p(m). A p(m), which deviates from the probability of a single measurement is incrementally updated with p(pn) of subsequent measurements. Therefore p(pn) of measurements are combined with the current cell probability p(m). The example shows some darker green cells which are cut by more than one ray. These cells are probably less occupied than lighter green cells, which are only intersected by a single ray. Figure 1b shows CUFD using the example of p2 and p3. The significance for occupied and free is set identical for an incremental update to show an immediate modification. p2 is added in M in Figure 1a. The cell mp2 is dedicated as occupied. In Figure 1b is added p3, which is not within mp2 , but the ray that connects OS with p3 intersects mp2 . This situation creates competing occupancy statements for mp2 . Thus, no unique cell occupancy is possible and the resulting declaration is unknown. Several intersections enable that initially occupied cells switch into free, although they are partially occupied by obstacles. Figure 1b shows this situation for mp2 up to mpn .

The phenomenon attributes to the aliasing effect [7]. The Cause is a rough discretization of analogous measurements and representation in a grid. Figure 2a and Figure 2b show the same situation as also Figure 1.a and Figure 1b but with a grid resolution dm/2. An increased grid resolution is comparable to an increased sampling rate which suppresses the aliasing effect. Figure 2b shows that each cell is cut only by one ray and no free-declaration is made by partial occupied cells. The observation in Figure 2b is only correct taking into account the used sensor resolution and the considered distance. For dissimilar terms, a doubling of the resolution does not have to be sufficient. An increasing of the grid resolution is not unlimitedly possible because a halving of the cell size leads to a cubic increasing of cell number by a constant expansion of M. An increase can lead to computational problems on limited hardware. The example in Figure 1 is simplified. The same significance is used for occupied and free modification of a cell. Practical approaches use the bayesian-probabilistic-update in general with different significances to update the grid conservatively. Each obstacle should be added to the grid with the first measurement. Cleaning out an obstacle from the grid should require several updates. This practice requires several overlapping rays to visualize CUFD at a prior occupied cell.

IOP Conf. Series: Materials Science and Engineering **261** (2017) 012013 doi:10.1088/1757-899X/261/1/012013

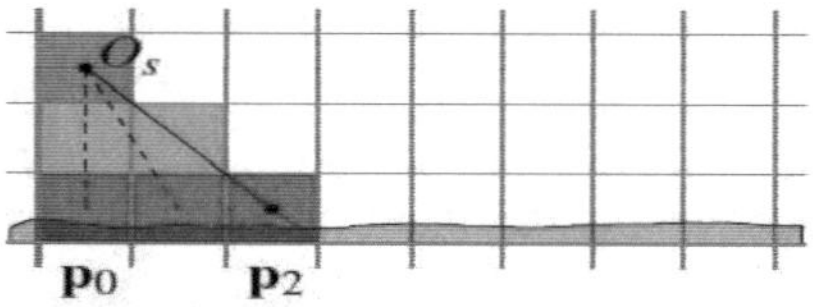

(a) Initial situation with a grid cell size d_m.

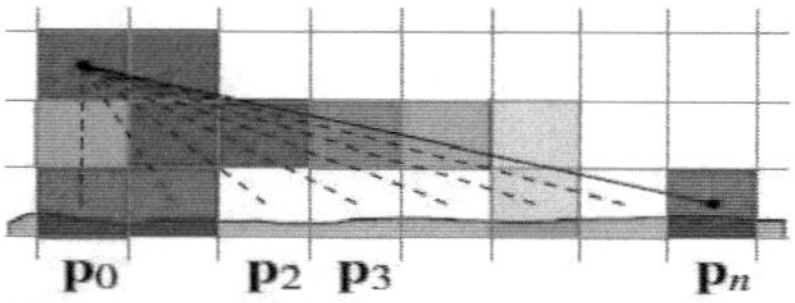

(b) Unwanted free-declaration at $\mathbf{p}_2$ up to $\mathbf{p}_n$ bases on 1a.

Figure 1: CUFD of occupied cells as Result of grid resolution d_m.

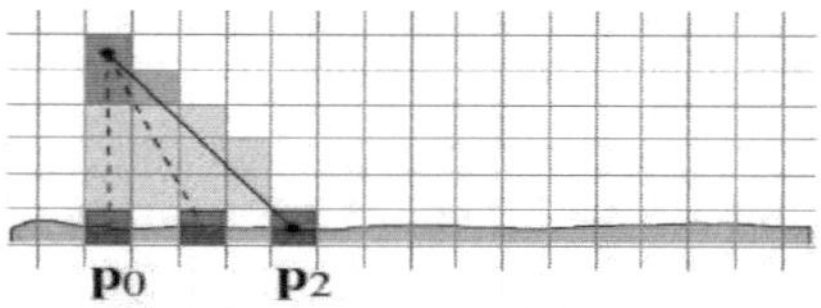

(a) Initial situation with grid cell size $d_m/2$.

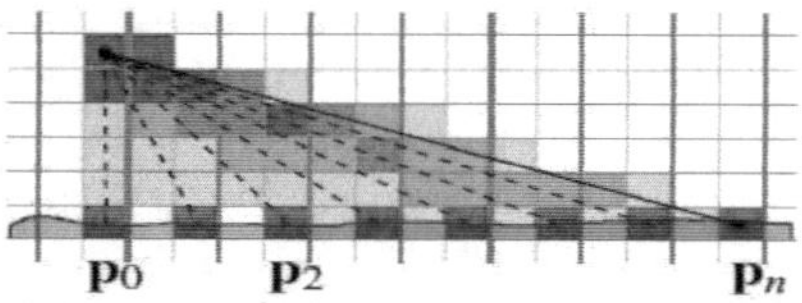

(b) Incremental update without unwanted free-declaration bases on 2a.

Figure 2: CUFD of occupied cells as Result of higher grid resolution $d_m/2$.

Figure 3 shows a real-environment situation. The data are generated during a flight of an unmanned helicopter through two pillars which are positioned on plane grassland, shown at Figure 3a. Figure 3b shows the used unmanned helicopter1. The data are generated with 2D Light Detection and Ranging (LiDAR) and are combined with the known pose of the helicopter to a 3D point cloud (Figure 3c). The expected result of a conversion of a recorded point cloud in Figure 3c to an occupancy grid is shown in Figure 3d. But Figure 3e shows the actual result when integrating all individual measurements sequentially into the grid. Significant parts of the grassland are not declared as occupied in the grid. For a better visualizing of the wrongly declared parts, the initial point cloud is depicted also. The incidence angles of the laser rays are almost parallel to the grassland plane ($\approx 70°$). As a result, a few cells representing the grassland are intersected by several rays. A greater number of overlapping rays against ending rays lead to CUFD and without further verification with the real situation a collision of the helicopter with the grassland could be possible.

(a) Onboard camera view at the pillows and grassland.

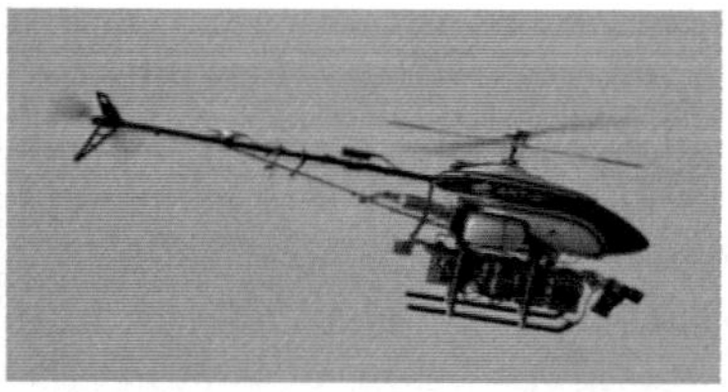
(b) Unmanned Helicopter ARTIS.

(c) Point cloud, generated with monogon LiDAR.

(d) Expected occupancy grid.

(e) Actually generated occupancy grid.

Figure 3: Example situation for CUFD.

The grid resolution should be based on the Nyquist-Shannon-sampling theorem to prevent the aliasing effect and CUFD. The grid resolution should be at least twice the sensor resolution at the coordinates at which a sensor measurement is made on an object surface. Such high grid resolution is not always predictable, especially onboard real-time approaches. The use of occupancy grids with a resolution that is less than the sensor resolution requires the development of approaches which reduce the effects of the aliasing effect. This means that the aliasing effect is not impeded, but it is less visible, as seen in Figure 3e.

3. Approaches

The current section presents three approaches to covering CUFD. The first approach is based on common literature and conducts as a comparison object. The second method works with an increase of the grid resolution based on a specific grid data structure. The last approach solves CUFD by considering grid cells not as independent spaces but rather by evaluating the cells based on their positions along the measuring ray.

3.1. Reduction of CUFD with large point clouds

Scherer [3] and Wurm [8] suggest as a solution of CUFD not to add single measurements to a grid but rather to add a pre-combined set of measurements Ci. This combination limits the mutual influence of neighboring measurements. Hornung [9] presents a FIFO buffering of several LiDAR scans St in a point cloud Ct−n:t=$\cup_{t-1:t}^{t} S_t$with n = const. For each measurement in Ct−n:t a ray and the respectively intersected cells in the grid are determined. The determined cells are uniquely stored in two separate sets (free, occupied). Subsequently, the elements of the sets are modified in the grid. Based on the assumption, the declaration significances of a one-time free- and occupancy-declaration are not the same. This procedure prevents a free-declaration of uniquely occupied cells for the measurements in Ci.

3.2. Covering CUFD with an octree

The use of a grid data structure that allows data to be extracted for subsequent computations in a reduced resolution from the grid as the resolution in which the data were added to the grid can also solve CUFD. The octree [10] data structure enables this function. If a measurement ray is added to an octree, leaves at the tree height h are declared with the respective occupancy probability. Updates of the inner tree nodes from the direct parent node to the root node follow the declaration of a leaf. In the current case, the update $\hat{p}(n)$ of an internal node is performed by $\hat{p}(n) = max_i P(n_i)$ on the basis of the maximum $p(n_i)$ of the child nodes. This procedure suppresses the free-declaration of nodes and the CUFD with each reduction of the tree depth d. Measurements are added to the tree at a tree depth $d_h = h - 1$. The measurements are obtained from the tree at level $d_{h-\lambda}$ with $\lambda \geq 1$. This approach works with an increase of the grid resolution with the drawback of a cubic growing cell number by halved resolution. Consequences are a higher memory consumption and higher computing requirements. In order to limit this drawback, only the additions of measurements in M at a high resolution ensure. Follow-up processes, such as the obstacle avoidance, extract the map at a lower resolution and therefore, no increasing memory and computing requirements are expected for these processes.

3.3. Conditional free-declaration of occupied cell

CUFD can also be prevented by the fact that the occupancy of a grid cell is not interpreted as an independent space description but rather is set in relation to its neighbors along a measurement ray. If the incidence angle of a ray Rp leading to a measurement p is relatively small, CUFD accrues among spatially close measurements along Rp between the first intersected occupied cell and the cell which contains the end of Rp. This condition occurs in two cases. Firstly, the gradient of the hit surface is locally almost parallel to the grid axes. The ray Rp runs from the first contact with an occupied cell in the interior of occupied cells. No free cells are cut. This situation represents the measurement p1 in Figure 4. Secondly, the gradient of the hit surface is locally non-parallel to the grid axes. This case eventuates the aliasing effect, Rp intersects recurrently occupied and free cells. Measurement p0 visualizes this case in Figure 4.

IOP Conf. Series: Materials Science and Engineering **261** (2017) 012013 doi:10.1088/1757-899X/261/1/012013

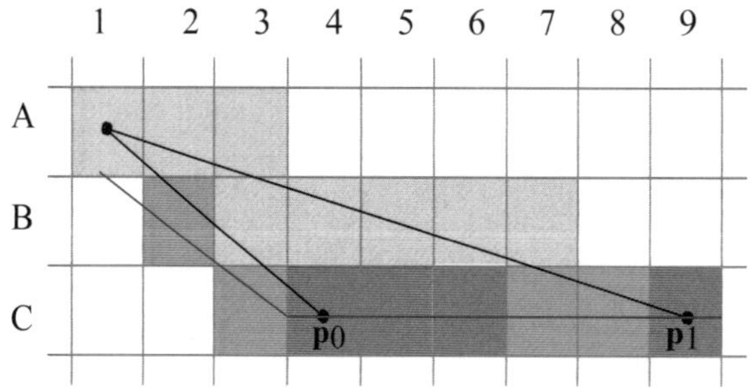

Figure 4: Ray path within the occupancy grid, which leads to CUFD.

$$S = \{q_0; q_1; q_2; q_3; q_4\}$$
$$\Sigma = \{occupied(p > p_{uk}); free(p \leq p_{uk})\}$$
$$\delta = \left\{ \begin{array}{ll} \sigma_0 & = \{p_m \leq p_{uk} \ \& \ i \leq th_{free}\}; \\ \sigma_1 & = \{p_m \leq p_{uk} \ \& \ i > th_{free}\}; \\ \sigma_2 & = \{p_m > p_{uk} \ \& \ j \leq th_{occ}\}; \\ \sigma_3 & = \{p_m > p_{uk} \ \& \ j > th_{occ}\}; \\ \sigma_4 & = \left\{p_m > p_{uk} \ \& \ m = \left\lfloor \frac{\mathbf{p}_i - \mathbf{t}_{mi}}{d_m} + 0.5 \right\rfloor \right\} \end{array} \right\}$$
$$F = \{q_3; q_4\}$$

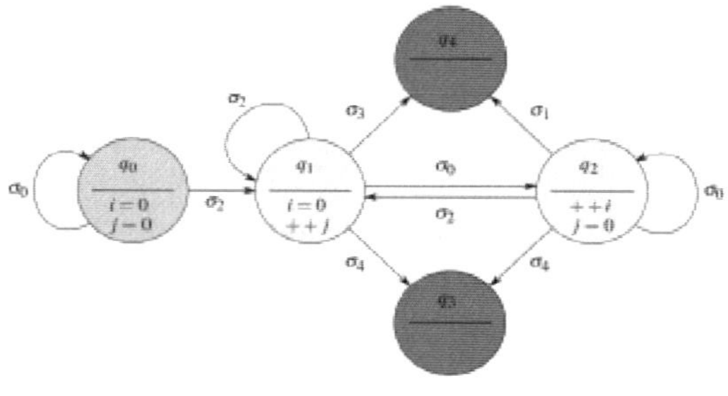

Figure 5: A finite-state machine for Conditional free-declaration along a ray Rp.

CUFD can be prevented in these cases by allowing a ray to intersect a number thocc occupied and thfree free cells in the neighborhood of p without reducing p(Z) of the intersected cells. This behavior requires that the cells in the neighborhood around p are not independent spatial descriptions but rather in relation to each other along Rp. The referencing of cells along Rp is carried out with a finite-state machine A = (S, Σ, δ , F, q0) runs in addition to the inverse sensor model for each cell. The parameters are: S set of states, δ state-transition functions, Σ input alphabet, F set of final states and q0 initial state. The behavior and the transitions of A are shown in Figure 5. The ray Rp starts at the sensor origin mOs and A is initialized at state q0. A remains at q0 to Rp intersects a cell mk with (p > puk). At q0 two counters i = 0 and j = 1 are initialized which control the thresholds thfree and thocc. thocc describes the maximum number of occupied cells in series along a ray, which could be intersected by Rp without modifying their occupancy probability. The second threshold thfree defines the maximum number of free cells between two occupied cells which are intersected by Rp without modifying their occupancy probability. If i and j are greater than the thresholds the current "the conditional free-declaration" approach is aborted for the current ray and is returned to the original approach. All measurements along Rp are updated independently of the relations between the cells with the original bayesian-probabilistic-update. A switches from q0 to q1 if Rp arrives the first occupied cell mocc,k, k = 1. The current occupancy probability of mocc,1 is not modified at q1. If further occupied cells mocc,k follow on mocc,1, A stays in q1 under the condition j ≤ thocc. If the condition j ≤ thocc is torn, A switches to the final state q4. The arrival of q4 prevents the conditional declaration for the current ray. If mocc,1 if followed by the end of the ray mp, A jumps in the final state q3. The arrival of q3 enables the conditional free-declaration. All cells intersected by Rp are not modified, except for mp which is updated by the bayesian-probabilistic-update. Follows on mocc,k a free cell mf ree,1, A switches from q1 to q2. This branch of A treats CUFD at an aliasing-situation. Follows on q2 further free cells mfree,k, A stays at q2 under the restriction i ≤ thfree. If the restriction breaks, A switches to q4 and it follows an update of all intersected cells. Is mfree,k; followed by an occupied cell with maintaining i ≤ thfree, A goes back to q1 and the branch to treat the aliasing effect is reinitialized. The reinitializing enables to deal with longer aliasing sections. If mfree,k if followed by the end of the treated ray mp, A is transmitted in the final state q3 and allows an execution of the conditional free-declaration. The ability to handle dynamic obstacles remains with this implementation.

4. Validation

The functionality of the three presented approaches is validated by a comparison against an expected nominal condition. The used data are generated with the LiDAR "Hokuyo UTM-30LX" carried by an unmanned helicopter. The expected nominal condition presents Figure 3d. The validation ensues in two steps. Firstly, several exemplary parameter sets are presented to get an idea which sets best prevent the CUFD. The used data contain three obstacles, two pillars and a ground plane. The pillars are clearly visible in Figure 3d as well as in Figure 3e. Otherwise, the ground shows a lot of holes base on CUFD. Therefore the aim is to identify sets which generate a closed ground plane. This validation is qualitative. The presented approaches are used in online mapping application for an Unmanned Aircraft (UA). Therefore the second validation focuses on run-time, complexity and memory consumption. This rating uses the best parameter set per approach of the first validation to get the processing performance for the best CUFD prevention.

4.1. Parameter set for the representation of a closed ground

The origin for parameter identification bases on an occupancy grid with a cell dimension d_m = 0:5m. The grid bases on an octree structure with a tree height h = 10. The result of pasting the example measurements into the grid shows Figure 6 which is a top view of the scene in Figure 3.

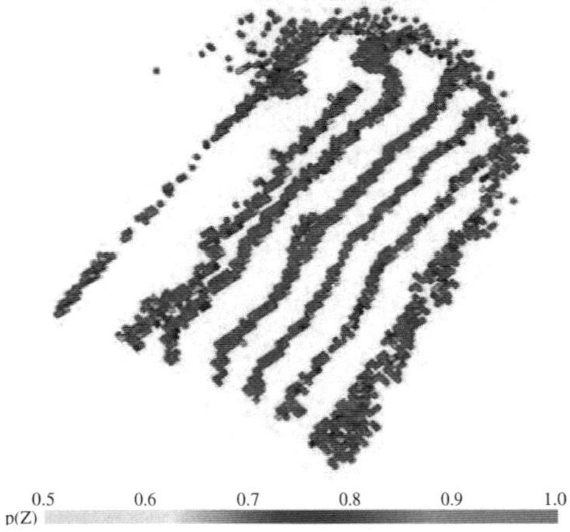

Figure 6: Original situation, without correction of CUFD. [11]

Table 1: Parameter sets to prevent CUFD.

approach	**3.1:** combine j scans S_i to a cloud $C = \cup_{i=0}^{j} S_i$	**3.2:** add at tree depth d_{in}, withdrawal at d_{out}, with $d_{out} < d_{in}$	**3.3:** conditional occupancy-declaration based on the thresholds th_{free} and th_{occ}
set 1	$j = 20$	$d_{in} = 12$ $d_{out} = 11$	$th_{free} = 1$ $th_{occ} = 1$
set 2	$j = 40$	$d_{in} = 12$ $d_{out} = 10$	$th_{free} = 4$ $th_{occ} = 1$
set 3	$j = 60$		$th_{free} = 1$ $th_{occ} = 4$
set 4	$j = 80$		$th_{free} = 2$ $th_{occ} = 8$

The coloring represents the occupancy probability p(Z) of each cell. The measurements of the used LiDAR are added in the grid with p(Z) = 0.8. Deviating probabilities hint to a usage of the bayesian-probabilistic-update. The figure shows only occupied cells with $0.5 \leq p(Z) \leq 1.0$. The grid shows that several parts of the ground plane are not visible and many cells have a $p(Z) \leq 0.8$. This observation indicates a contrary occupancy interpretation of some cells. A major cause is next to sensor noise the CUFD. The approaches are intended to close these holes and covers CUFD. To enables this aim, a choice of parameter sets for each approach is shown in table 1.

Results to Approach 3.1 Figure 7 presents the results of approach 3.1, for each subfigure j scans are combined to point cloud Ci and integrated into the grid. The increase from j = 20 in Figure 7a to j = 80 in Figure 7d reduces CUFD and increases the number of cells with p(Z) > 0.8. But even Figure 7d shows holes and therefore a complete prevention of CUFD cannot be presented.

Results to Approach 3.2 Figure 8 shows the use of this approach by adding measurements at tree depth d = 11 or d = 12 and extracting at d = 10. Figure 8a shows an increased input depth about one level which reduces CUFD. An increase about a further level of the input depth covers CUFD (Figure 8b) completely.

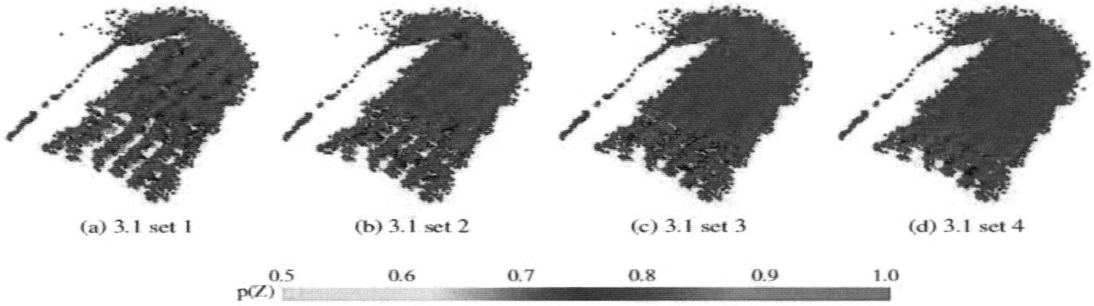

Figure 7. Parameter comparison for Approach 3.1.

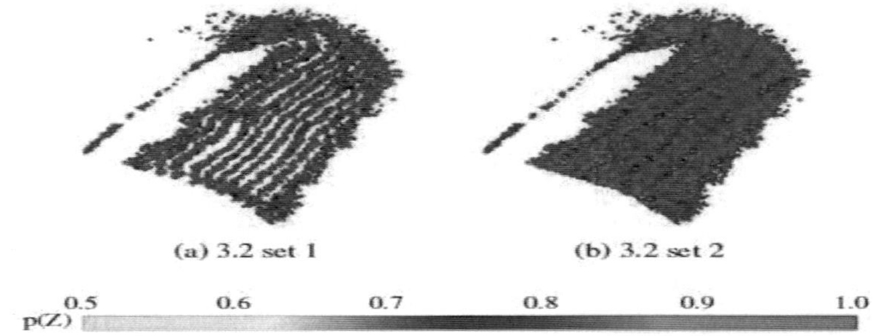

Figure 8. Parameter comparison for Approach 3.2.

Results to Approach 3.3 Figure 9 shows the results of approach 3.3. Figure 9a and Figure 9b show marginal changes against Figure 6. An increase of th_{occ} in 3.3(3) leads to a reduction of CUFD. Further adaptions of the thresholds lead to a complete prevention of CUFD and fewer cells with $p(Z) \leq 0.8$.

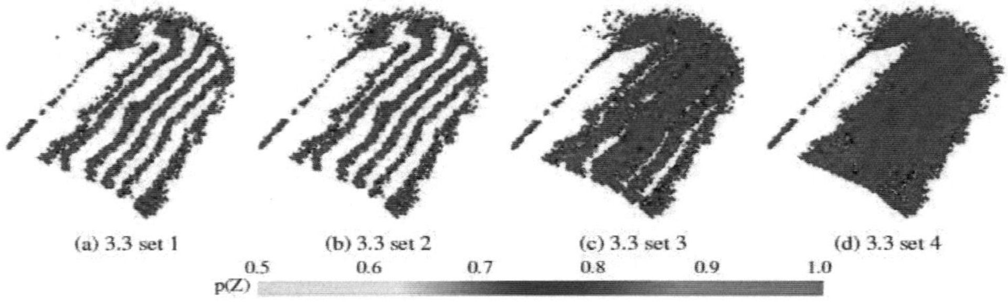

Figure 9. Parameter comparison for Approach 3.3

The results of the comparison show that all approaches are useable to more or less prevent CUFD at the ground plane for the used example. In a visual comparison, the parameter sets 3.1(4), 3.2(2) and 3.3(4) show the best results. These sets are used in the next section to analyze the performances between the three approaches.

4.2. *Performance analysis*

The presented approaches are used in an online application by UA and therefore have to be computable on limited hardware (Proc.: i5-4210, RAM: 4GB) onboard a vehicle. A comparison of the parameter sets 3.1(4), 3.2(2) and 3.3(4) should show which approach is best suited to run on limited hardware. Therefore the runtime per added point cloud, the number of iterations in the octree per point cloud and memory requirements of the grid are collected. Figure 10c shows the results. The regarded values are logged over recording time and bases on the shown real LiDAR data. A comparison between the approaches for the number of insertions in the octree per point cloud is presenting in Figure 10a. Approach 3.1(4) deviates from 3.2(2) or 3.3(4). The deviation shows a magnitude of about eighty times the results of 3.2(2) per

IOP Conf. Series: Materials Science and Engineering **261** (2017) 012013 doi:10.1088/1757-899X/261/1/012013

step. This observation bases on the combination of j = 80 scans to a point cloud at 3.1(4) which is repeatedly added. The number of insertions at 3.1(4) also leads to an increased runtime per added point cloud into the octree, shown in Figure 10b. The differences are less because single measurements of several scans are pre-combined before they are added to the tree. The combination of measurements reduces the deviations between the approaches, but the great number of insertions at 3.1(4) cannot be balanced. The number of insertions is approximately equal between 3.2(2) and 3.3(4). This observation bases on the procedure, that both approaches add measurements directly into the octree without further pre-processing.

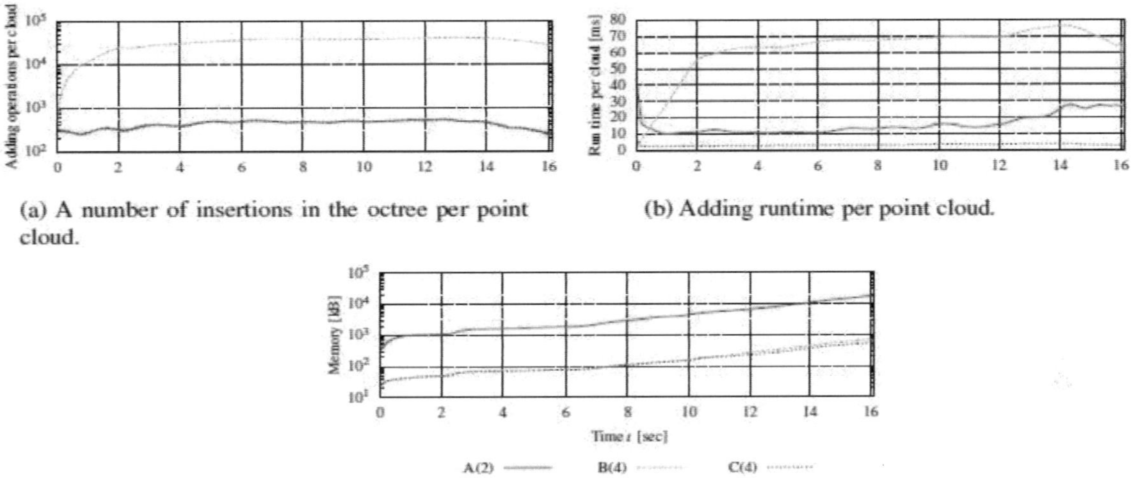

(a) A number of insertions in the octree per point cloud.

(b) Adding runtime per point cloud.

(c) Memory requirements of the octree.

Figure 10. Performance comparison of 3.1(4), 3.2(2) and 3.3(4). [11]

The runtime per added point cloud shows deviations between 3.2(2) and 3.3(4). These differences base on the size of the respective octree and on the number of constitutive operations like the number of single cell operations for each added ray. Some operations in an octree such as traversing (Θ(n)) have an asymptotic behavior and depend on the number of nodes n at the tree. An increased tree height h leads to an increased number of nodes (cells), nh=12 = N2 ∗ nh=10, whereby N = 8 is the number of children per node. This property leads to increased processing effort and runtime for trees with a taller height. An example is traversing trough inner nodes after updating a cell declaration. A further cause of increased runtime is a higher cell resolution at 3.2(2). This leads to a situation that fewer measurements are spatially located in the same cell. The modification of a cell often bases on a single measurement per cell. An insertion of a point cloud into a higher resolved tree results in a major number of occupied cells. Additionally, based on the reduced cell size, a ray intersects more cells in a higher resolved tree and therefore more free cells. Both points lead to increased number of touched cells and more subsequent operations which lead to an increased runtime at 3.2(2).

The memory requirements of a cell are constant and independent of its depth in the tree. An increased tree height leads to an increased number of inner nodes. Further, each added measurement touches more cells in a higher resolved tree than in a lower resoled tree. Both points are connected in 3.2(3) and lead to higher memory requirements against the other presented approaches, as seen in Fig 10c. Approach 3.1(4) has higher memory requirements compared to 3.3(4), which cannot be explained by a different tree height. At the end (16.15 sec) of the scenario, the number of the occupied cells is nocc = 4880 for 3.1(4) and nocc = 4895 for 3.3(4). This deviation cannot be the cause of different memory requirements. The cause is the number of free cells n free = 20957 for 3.1(4) and n free = 14590 for 3.3(4), at the end of the logged test. The difference between the numbers of the free cells is attributable to properties of the added point clouds. Each cloud has a single sensor position per cloud for all included measurements. For the transformation of single measurements from sensor coordinates into global world coordinates and combination into a point cloud, a compensation of the sensor motions is necessary. This procedure requires that for each measurement the correct global sensor position is used at the moment of recording, especially if a system like a rotating LiDAR is used. After this procedure, each point cloud often gets only one single sensor position for a roughly global assignment. Commonly the sensor position of the first or the last added measurement is used. This characteristic is also used in the implementation described above. An insertion of a point cloud into a grid requires a conversion of contained measurements into rays, which based on measurements and the overall sensor position of the cloud. Therefore, the generated rays are not identical with the actually measured rays. This adaption of a ray leads to the situation that measurements theirself stay at the same position and the number of occupied cells does not change. But trace and length of a ray change and lead to an adjusted number of free cells and changed cell positions compared to the original rays. A rudimentary example of this description is depicted in Figure 11. The continuous black rays connect real sensor positions with the corresponding measurements.

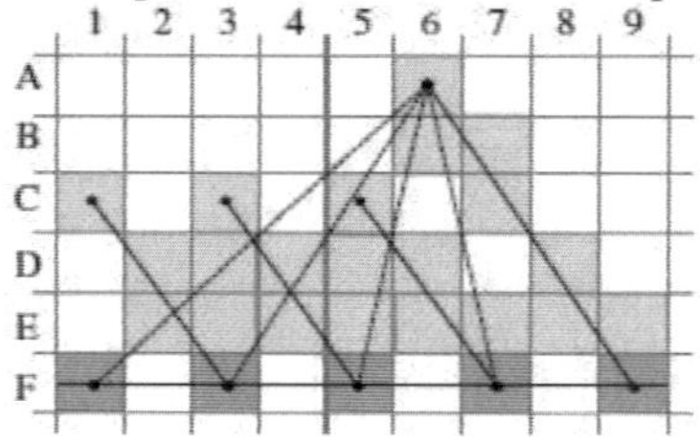

Figure 11. Adding single measurement rays against adding a cloud with one origin.

Cells which contain a measurement are declared occupied and colored red. All intersected cells with no measurements are colored green and declared as free. The overall point cloud contains the same example measurements but uses only one sensor position, in this example the position at cell (A6). Based on this position will be defined occupied and free cells with dashed rays. The declaration of occupied cells shows no difference between adding measurements continuously and adding the overall point cloud at once. But the declaration of free cells show some differences when looking at B5, C4, C6 and D7 which are intersected only by dashed rays. These intersections declare cells as free without real measurement support for this classification. This behavior is critical and can be the cause of collisions. The example describes a combination of single measurements to a point cloud. Approach 3.1(4) combines several point clouds to a major point cloud. This procedure leads to a much heavier influence of this phenomenon. Therefore the described behavior is assumed to be a cause for the difference between the memory requirement of 3.1(4) and 3.3(4). The qualitative and quantitative comparison between the approaches 3.1, 3.2 and 3.3 and especially the parameter sets 3.1(4), 3.2(2) and 3.3(4) show that 3.3 generates the best results to prevent CUFD during mapping LiDAR data for the presented flight scenario. This statement bases on less runtime, memory requirements and correct visualization of the mapped environment.

5. Conclusion

This paper deals with the phenomena of the conditional unwanted free-declaration (CUFD) which can occur during insertion of continuous distance measurements into a discrete probability occupancy grid. The cause of CUFD is a low spatial resolution of an occupancy grid compared to continuous distance measurements which are added into the grid and the resulting possibility that several measurement rays intersect a single grid cell. If this cell is only partly occupied by an obstacle, opposing occupancy statements (free/occupied) are generated. Contrary occupancy statements are generally desirable at probability occupancy grids to handle sensor noise or dynamic obstacles. But CUFD can also accrue at static objects and leads to a free-declaration of partly occupied cells. Free-declaration of occupied spaces in an environment representation which is used for unmanned obstacle avoidance is a danger because it can result in a collision with a non-mapped obstacle.

The aim is to reduce the differences between a mapping representation and the real world, to prevent possible collisions. Therefore, three approaches are presented to suppress CUFD. These approaches are validated qualitatively and quantitatively to compare the prevention of CUFD and the ability to process on limit hardware.

As result an approach is preferred which interprets cells along a measurement ray not as independent spaces rather the occupancy probability of a cell is set in relation to its neighbors. This procedure allows handling CUFD with less processing performance compared to other approaches.

References

[1] F. Andert and L. Goormann. "Combined grid and feature-based occupancy map building in large outdoor environments". In: *Intelligent Robots and Systems, 2007. IROS 2007. IEEE/RSJ International Conference on*. 2007, pp. 2065–2070. ISBN: 978-1-4244-0912-9.

[2] S. Hrabar and G. Sukhatme. "Vision-based navigation through urban canyons". In: Journal of *Field Robotics* 26.5 (2009), pp. 431–452.

[3] S. Scherer. "Low-Altitude Operation of Unmanned Rotorcraft". PhD thesis. Pittsburgh and PA: The Robotics Institute, Carnegie Mellon University, 2011.

[4] H. P. Moravec and A. Elfes. "High resolution maps from wide angle sonar". In: *Robotics and Automation. Proceedings. 1985 IEEE International Conference on*. Vol. 2. 1985, pp. 116–121.

[5] A. Elfes. "Using occupancy grids for mobile robot perception and navigation". In: *Computer* 22.6 (1989), pp. 46–57.

[6] U. Raschke and J. Borenstein. "A comparison of grid-type map-building techniques by index of performance". In:*Robotics and Automation, 1990. Proceedings., 1990 IEEE International Conference on*. 1990, pp. 1828–1832. ISBN: 0-8186-9061-5.

[7] E. Angel. *Interactive computer graphics: A top-down approach using OpenGL@*. 4th ed., internat. ed., [Nachdr.] Boston and Mass: Pearson/ Addison-Wesley, 2006. ISBN: 032131252X.

[8] K. M. Wurm et al. "OctoMap: A probabilistic, flexible, and compact 3D map representation for robotic systems". In: *Proc. of the ICRA 2010 workshop on best practice in 3D perception and modeling for mobile manipulation* (2010).

[9] A. Hornung et al. "OctoMap: an efficient probabilistic 3D mapping framework based on octrees". In: *Autonomous robots* 34.3 (2013), pp. 189–206.

[10] D. Meagher. *Geometric modeling using octree encoding*. Vol. IPL-TR-81-005. Reports / Rensselaer Polytechnic Inst Troy NY Image Processing LAB.

[11] S. Krause. "Sichtfelderweiterung eines Laserscanners mittels einer Kamera fuer unbemannte Luftfahrzeuge". PhD thesis. Braunschweig: Technischen Universitt Carolo-Wilhelmina zu Braunschweig, 2016.

IOP Conf. Series: Materials Science and Engineering **261** (2017) 012014 doi:10.1088/1757-899X/261/1/012014

Underwater 3D Reconstruction Based on Geometric Transformation of Sonar and Depth Information

Mingjie Dong[1,*], Wusheng Chou[1] and Guodong Yao[1]

[1]The State Key Laboratory of Virtual Reality Technology and Systems, Beihang University, XueYuan Road No.37, HaiDian District, Beijing, 100191, P.R.China

[*]Corresponding author: dongmj@buaa.edu.cn

Abstract. 3D reconstruction is of vital importance to detect and monitor the underwater environment. A method based on geometric transformation of mechanical scanning sonar and depth information is proposed, in which the point cloud data from sonar and depth gauge are acquired to reconstruct the underwater 3D environment. However, noise and interference can affect the measurement of sonar, and movement of sonar during measurement can lead to distortion of the received data. Meanwhile, translation and rotation movement of sonar head may happen when ROV dives which can lead to different body reference coordinates of different scanning. To solve this, pre-processing and motion compensation are implemented at first, and underwater matching correction algorithm is used to calculate the translation and rotation of the sonar head. Then the inverse operation is implemented to convert the scan data of every depth into the same coordinate reference system. Finally, surface reconstruction of point clouds from sonar the depth information are used to reconstruct underwater environment based on MLS (Moving Least Square Method) using PCL (Point Cloud Library). Water tank experiments verify the effectiveness of the proposed method.

1. Introduction

There is increasing interest in underwater imaging and inspection in recent years. The most common technique adopted to complete underwater 3D reconstruction is based on optical cameras and sonar devices. However, the measurement ranges of the optical devices are rather limited without external sources of light and almost non-existent with high turbidity, and these make acoustic imaging the most reliable means to obtain information from underwater environment [1].

There has been many 3D reconstruction techniques using all kinds of sonar. In [2], a novel model-based approach for 3D underwater scene reconstruction is proposed using Side Scan Sonar (SSS) arrays in complex and highly reverberating environments, while in [3], shape-from shading technique is adopted to process reconstruction using side scan sonar arrays. Beyond that, Synthetic Aperture Sonar (SAS) techniques also have interesting results in [4]. Nevertheless, SSS and SAS are all too expensive to be widely used. However, the underwater environment we need to reconstruct is simple, so, we use a mechanical scanning sonar and reduce the influence of vertical aperture through choosing the maximum value of the returned echo intensity as the obstacle in horizontal plane as vertical aperture zero.

As for 3D reconstruction, there has already been many studies. In [5], the surface is reconstructed through the use of computational geometry techniques. Paper [6] presents a practical mesh method to achieve an accurate reconstruction of seabed surface from raw sonar records. The method we use to

IOP Conf. Series: Materials Science and Engineering **261** (2017) 012014 doi:10.1088/1757-899X/261/1/012014

reconstruct the underwater environment is MLS based on PCL, which is a comprehensive free, BSD licensed, library for n-D Point Clouds and 3D geometry processing [7].

2. Processing of point cloud data from depth gauge and sonar

2.1. Acquirement of point cloud data

A depth gauge is a pressure gauge that displays the equivalent depth in water, which is well known and will not be described in detail here. And the sonar we use to reconstruct the underwater environment is a Tritech Micro DST sonar, which can be programmed to cover variable length sectors from a few degrees to full 360° scans with an operating range of up to 75 meters.

The sonar performs scans in a horizontal 2D plane by rotating a mechanically actuated transducer head at pre-set angular increments [8]. For each emitted beam, an echo intensity profile is returned from the environment and discretized into a set of bins [9]. However, in a practical application, every ping received will contain noise and interference except for useful information, especially within the limited space like a water tank. Meanwhile, the ROV's motion can induce a distortion in the acoustic image when ROV moves, which is very common in water. So, it is necessary to compensate the distortion induced by motion and pre-process the sonar data before reconstructing the underwater environment. The method we use to compensate the motion induced distortion and pre-process the sonar data are shown in our previous paper [10]. The result of the acquirement with one sonar scan after underwater matching correction algorithm is as in figure.1.

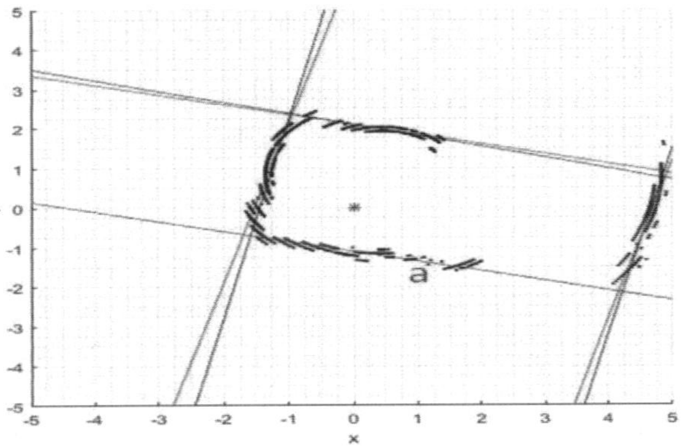

Figure 1. Point cloud data of one sonar scan after correction.

2.2. Transformation of Point Cloud Data

Let's define each scan as $S_i(i=1,2,\cdots n)$, where n means the nth scan at the nth depth. At the same time, we set the first scan as the reference with other scans transformed to the same coordinate system of the first scan $S_1 = S_{ref} = \{q_1, q_2, \cdots q_n\}$, where q_i represents the ith measurement in a scan. For each scan, the position of the sonar head is calculated using the geometric matching correction algorithm to measure the relative translation displacement of two adjacent measurements based on the position the sonar head relative to the base reference coordinate system, meanwhile, the slope of the base line in two adjacent scans at different depth is used to calculate the rotation angle. The reason for choosing the base line in two adjacent scans is that the change between two adjacent scans is very small and easily recognized. When we calculate the translation displacement and rotation angle of two adjacent scans at two adjacent depth, we can deduce the relative translation displacement and rotation angle using recursive computation.

For two adjacent depth as shown in figure.2, the position of the sonar head relative to origin of the base reference system is calculated using geometric matching correction algorithm, as in figure.3.

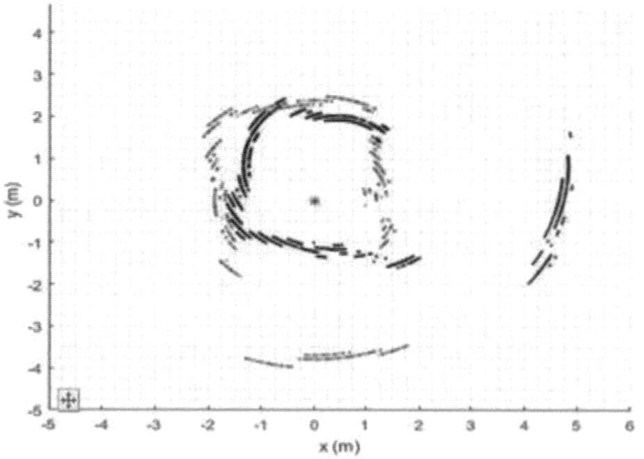

Figure 2. Example of two adjacent scans.

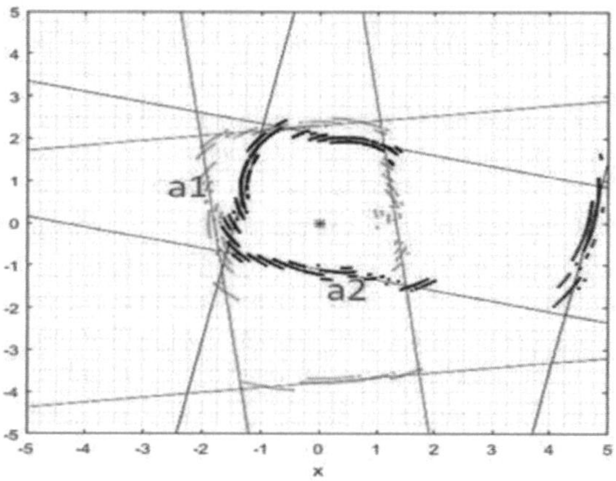

Figure 3. Geometric matching correction of the two adjacent scans

Let's assume that the scan with a1 inside is the s_{ref}, the scan with a2 inside is the second scan, then, the translation the second scan relative to the last scan can be calculated using the algorithm above, which can be defined as $\left[T_x, T_y\right]$. The relative rotation angle, which is also the angle of a1 and a2, can be calculated using equation (1) with k_1 and k_2 the slopes of the straight line a1 and a2 with counter clockwise direction positive.

$$\theta = \arctan\left(\frac{k_2 - k_1}{1 + k_1 k_2}\right) \tag{1}$$

After knowing the translation displacement and rotation angle, we can get equation (2),

$$\begin{bmatrix} x_2 & y_2 & 1 \end{bmatrix} = \begin{bmatrix} x_1 & y_1 & 1 \end{bmatrix} \bullet \begin{bmatrix} \cos\theta & \sin\theta & 0 \\ -\sin\theta & \cos\theta & 0 \\ 0 & 0 & 1 \end{bmatrix} \bullet \begin{bmatrix} 1 & 0 & 0 \\ 0 & 1 & 0 \\ T_x & -T_y & 1 \end{bmatrix} \tag{2}$$

The equation (2) can also be simplified as follows,

$$S_2 = S_1 \bullet T_1 \bullet R_1 \tag{3}$$

Where, T and R represents the corresponding translation matrix and rotation matrix, respectively. Then all the second scan points $S_2 = \{q_1, q_2, \cdots q_n\}$ can be transformed to the same coordinate system as the first scan. Similarly, we can calculate each scan's translation displacement and rotation angle relative to the last scan, and transform all the other scans to the same coordinate system as the first scan, just as equation (4).

$$S_i = S_1 \bullet T_1 \bullet R_1 \bullet T_2 \bullet R_2 \bullet \cdots T_{i-1} \bullet R_{i-1} (i = 2, 3 \cdots n) \tag{4}$$

With all the measurements scanned at different depths transformed to the coordinate system of the first scan, we can process the surface reconstruction of underwater environment with the point cloud data from sonar and depth information.

3. Surface reconstruction of point clouds data

For the point cloud data we collect is discrete and irregular, we should preprocess the point cloud at first before surface reconstruction.

3.1. Outlier removal

Radius outlier removal filter is implemented to remove the outliers. The principle is as shown in figure.4. At first, we set each point as the centre of a circle with radius R, respectively. Then, we set the minimum neighbours represented using N within R. If the numbers of the neighbours within R is less than N, then the point is considered as the outlier and will be removed. The values of the R and N depend on the circumstances.

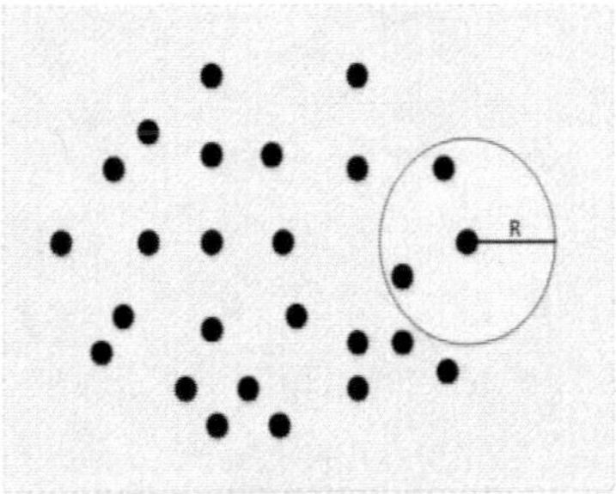

Figure 4. Radius outlier removal filter.

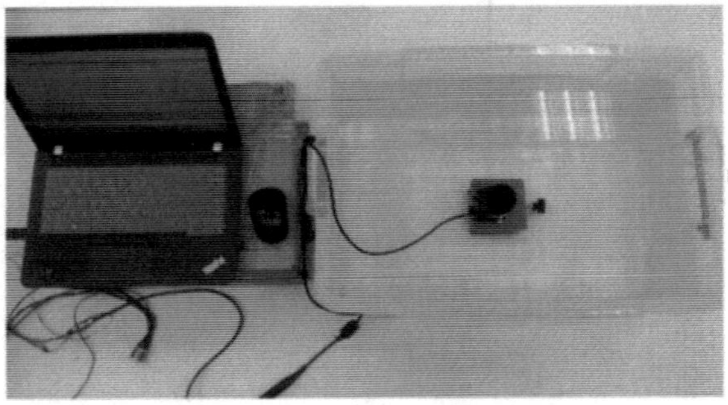

Figure 5. Water tank experimental platform.

3.2. Surface fitting based on MLS

As we all know, the cloud data can't be connected together simply to process surface fitting, especially when it is discrete. If we know the style of the surface, we can use least square method to get the fitting surface by solving linear equation sets. However, the least square method won't work if the discrete point cloud data is larger and the style of the surface is unknown and complex. In this situation, the MLS is very suitable for surface fitting. During the processing of MLS, the fitting function is defined in the local sub-domain of the fitting area, just as equation (5).

$$f(x)=\sum_{i=1}^{m}\alpha_i(x)p_i(x)=P^T(x)\alpha(x) \tag{5}$$

Where, $\alpha(x)=[\alpha_1(x),\alpha_2(x),\cdots\alpha_m(x)]^T$ is function of spatial coordinate x, $P(x)=[p_1(x),p_2(x),\cdots p_m(x)]^T$ is the basis function, with m the number of the basis function. As an example, if it is linear basis function for two dimensional problems, then,

$$P(x)=[1,x,y]^T, m=3 \tag{6}$$

Or else, it is the quadratic basis function, then,

$$P(x)=\left[1,x,y,x^2,xy,y^2\right]^T, m=6 \tag{7}$$

So, in order to get the optimal fitting surface, we should get the optimal $\alpha(x)$, which can be obtained by obtaining the minimum value of the weighted discrete L2-norm as shown in equation (8).

$$J=\sum_{i=1}^{n}\omega_i\left[P^T(x_i)\alpha(x)-y_i\right]^2 \tag{8}$$

Where, n represents the number of nodes in the region, y_i represents the value of the node x_i, and ω_i is the weight function of the node x_i. The weight function in the MLS should have the characteristic of compact support, which means that the value of weight function in a sub-domain of x is nonzero while the value is zero outside of the sub-domain. Besides, the weight function should be nonnegative, and should be monotonically decreasing with the increase of x. The most commonly used weight function is a spline function. To get the optimal $\alpha(x)$, we should deduce the minimum value of equation (8), as shown in equation (9).

$$\frac{\partial J}{\partial \alpha}=0 \tag{9}$$

Then, we can get,

$$\sum_{i=1}^{n}\omega_i p(x_i)p^T(x_i)\alpha(x)-\sum_{i=1}^{n}\omega_i p^T(x_i)y_i=0 \tag{10}$$

Let's assume that,

$$A(x)=\sum_{i=1}^{n}\omega_i p(x_i)p^T(x_i) \tag{11}$$

$$B(x)=\sum_{i=1}^{n}\omega_i p^T(x_i) \tag{12}$$

$$y^T=[y_1,y_2,\cdots y_n] \tag{13}$$

Then, we can get,

$$A(x)\alpha(x)-B(x)y=0 \tag{14}$$

$$\alpha(x)=A^{-1}(x)B(x)y \tag{15}$$

Finally, we can deduce the solution of the MLS with equation (15) substituted into equation (5).

$$f(x)=p^T(x)A^{-1}(x)B(x)y \tag{16}$$

After figuring out the principles of the MLS, we can easily programme in the PCL using its function library, to get the fitting surface using the point cloud data we get.

4. Water tank experiment

In order to verify the effectiveness of the 3D underwater reconstruction algorithm, we build a test platform to get the absolute value of the depth information and the environment information for the lack of suitable underwater environment.

4.1. Experiment test platform of the underwater 3D reconstruction

The water tank we used is as in Figure.5, which is composed of mechanically scanned sonar and a lifting platform. The size of the water tank is 580mm×380mm×280mm (length×width×depth), with water depth 212mm during the experiments. The range of the lifting platform is from 46mm to 150mm, which means the height the upper surface of the lifting platform relative to the ground surface.

During the experiment, the height of the sonar to the bottom of the water tank is set up manually using micrometre to measure, which substitutes for the depth gauge of the ROV equivalently. The direction of the sonar head is different each time of the scanning, which needs to be transformed using geometric transformation. A total of 33 set of data the sonar relative to the bottom of the water tank is collected, which is shown as in figure.6.

4.2. Results of the transformation of point cloud data from sonar

After collecting the 33 set of sonar data during different depth, the underwater matching correction algorithm is used to calculate the translation and rotation of the sonar. Then the inverse operation is implemented to obtain the scan data of every depth in the same coordinate reference system. The results of the transformation of point cloud data from the mechanically scanned sonar is as in figure.7

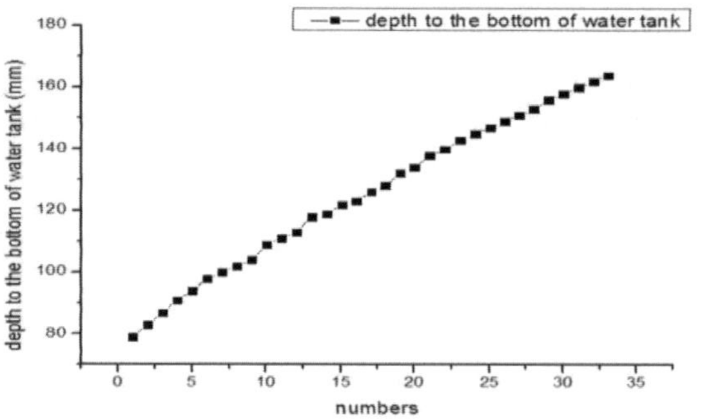

Figure 6. Height of the sonar head relative to the bottom of water tank.

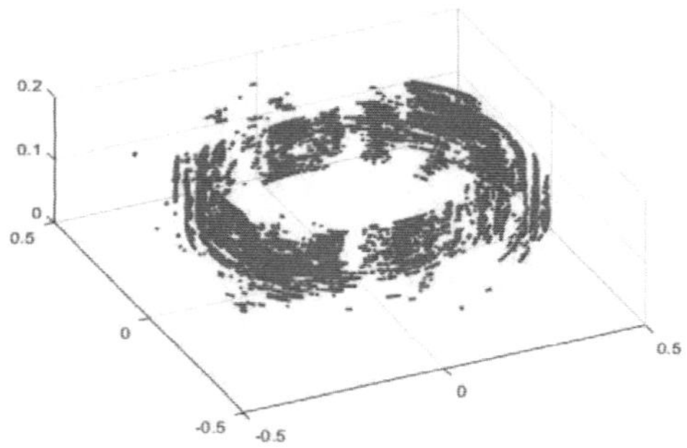

Figure 7. Results of point cloud data transformation from sonar.

Then we remove the outliers using radius outlier removal filter, with R=0.02, N=50 . After the preliminary processing of the sonar data using PCL, the underwater environment after reconstructing is as in figure.8.

Finally, we process the MLS to fit the surface of the point cloud data, finishing the underwater environment reconstruction, the result is as shown in figure.9.

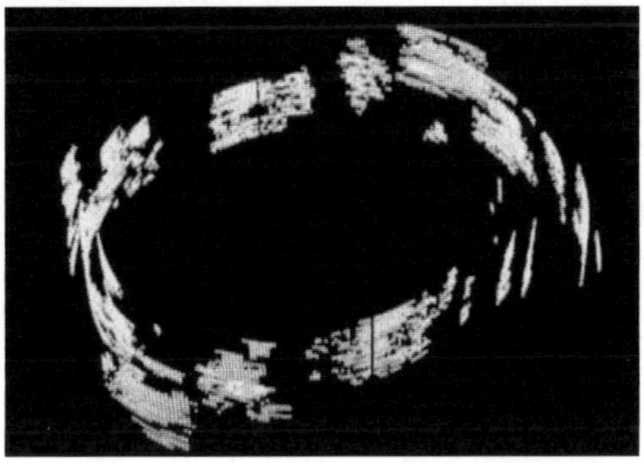

Figure 8. Preliminary processing using PCL

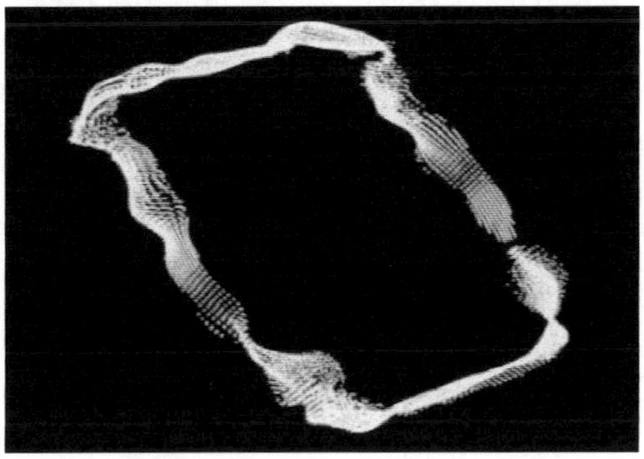

Figure 9. Result of the surface reconstruction.

5. Conclusions

The paper proposes a method to complete underwater environment 3D reconstruction based on point cloud data from mechanical scanning sonar and depth information using PCL. At first, pre-processing and motion compensation are implemented to remove noise and other interference. After that, underwater matching correction algorithm is used to calculate the translation and rotation of the sonar head. Inverse operation is implemented to convert the scan data of every depth into the same coordinate reference system in the next moment. Finally, surface reconstruction of point clouds from sonar the depth information are used to reconstruct underwater environment based on MLS using PCL. Water tank experiments verify the effectiveness of 3D reconstruction method, paving the way for future more precise 3D underwater reconstruction applications.

Acknowledgments

The authors would like to thank the support of National Natural Science Foundation of China under grant No.61633002, and the Natural National Key Basic Research Program of China under grant No.2013CB035503.

References

[1] Brahim N, Gueriot D, Daniel S, et al. 3D reconstruction of underwater scenes using DIDSON acoustic sonar image sequences through evolutionary algorithms [C]. Oceans. IEEE, 2011, 1-6.

[2] Saucan A A, Sintes C, Chonavel T, et al. Model-Based Adaptive 3D Sonar Reconstruction in Reverberating Environments [J]. IEEE Transactions on Image Processing A Publication of the IEEE Signal Processing Society, 2015, 24(10):2928-40.

[3] Robleskelly A, Hancock E R. Shape-from-shading using the heat equation [J]. IEEE Transactions on Image Processing a Publication of the IEEE Signal Processing Society, 2007, 16(1):7-21.

[4] Coiras E, Groen J. 3D target shape from SAS images based on a deformable mesh [C]. Proceedings of 3rd International Conference on Underwater Acoustic Measurements, 2009:21-26.

[5] Ouyang D, Feng H Y. Reconstruction of 2D polygonal curves and 3D triangular surfaces via clustering of Delaunay circles/spheres [J]. Computer-Aided Design, 2011, 43(8):839-847.

[6] Jouvencel B, Xiang X, Zhang L, et al. 3D Reconstruction of seabed surface through sonar data of AUVs [J]. Indian Journal of Geo-Marine Sciences (IJMS), 2012, 41(6): 509-515.
[7] Rusu R B, Cousins S. 3D is here: Point cloud library[C]. IEEE International Conference on Robotics & Automation. 2011:1-4.
[8] Ribas D, Ridao P, Neira J. Underwater SLAM for Structured Environments Using an Imaging Sonar [M]. Springer Berlin Heidelberg, 2010.
[9] Ribas D, Ridao P, Tard, et al. Underwater SLAM in man-made structured environments [J]. Journal of Field Robotics, 2008, 25(11-12):898–921.
[10] Dong M, Chou W, Fang B. Underwater Matching Correction Navigation Based on Geometric Features Using Sonar Point Cloud Data[J]. Scientific Programming, 2017, 2017.

IOP Conf. Series: Materials Science and Engineering **261** (2017) 012015 doi:10.1088/1757-899X/261/1/012015

Nonlinear Robust Controller Design for Plasma Magnetic Control in Fusion Reactors

C Tunceroglu[1,a] and U Hasirci[2,b]

[1]Department of Electrical and Electronics Eng., Duzce University, Duzce, 81620, Turkey
[2]Department of Electrical and Electronics Eng., Duzce University, Duzce, 81620, Turkey

[a] cagdastunceroglu@duzce.edu.tr; [b]ugurhasirci@duzce.edu.tr

Abstract. Tokamak is a fusion reactor that provides extremely high temperature needed for fusion process. Such a temperature is created by confining plasma via some poloidal coils. Control of voltages to be applied to the poloidal coils is an important control problem for proper operation of the overall system. This paper deals with the nonlinear robust control of plasma current, shape, and position in a Tokamak reactor. The paper briefly discusses the modeling issue of a Tokamak, and then, based on a nonlinear state-space model, presents three different versions of a robust controller for the process. All parameters of the system are assumed to be unknown for all three types of the controllers. Lyapunov-type arguments are used to prove the stability of the proposed control schemes. Advantages and disadvantages with each type of controllers are discussed in detail.

1. Introduction

Tokamak is a very powerful reactor to produce a very high temperature, approximately 100×106 °C [1], needed for fusion process. Main function of this reactor is to confine the plasma by using some poloidal coils. Controlled voltages are applied to poloidal coils and this creates currents that produce the magnetic field. The mutual interaction between the currents and magnetic fields ensures that the plasma is positioned where heating is more efficient [2]. Well-known illustration of a Tokamak reactor, illustrated by European Fusion Development Agreement (EFDA) can be found in [3].

Feedback control is used to adjust the voltages of poloidal coils and, consequently, plasma current and plasma distance to wall. A designer has three options for designing a feedback controller for the system; (i) Linear controller, (ii) Nonlinear controller with exact model knowledge, and (iii) Nonlinear controller with parameter uncertainty.

Dynamic model of a Tokamak is highly nonlinear. A linear controller needs a linearized form of this model. But stability result of a linear controller for a nonlinear system is valid only at a small neighborhood of the equilibrium points of the states. Moreover, linearization may cancel some useful nonlinearities such as having quenching effects ones. On the other hand, linear control has a mature literature and many successful tools are available for designing a linear controller. For example, Belyakov and Kavin produces a linear model for control purposes in [4]. In [5], a very simple and control-oriented model is derived, and a linear robust controller is presented in [6]. Some other linear controllers for various types of Tokamak devices (JET, DIII-D, JT-60U) can be found in [7-14].

Nonlinear controllers, on the other hand, does not use any linearization method and, for this reason, do not have the drawbacks mentioned above. A failure in taking into account of any part of such a complex fusion reactor dynamics may lead a serious damage to the devices [15]. Designing and implementing a nonlinear controller is a good solution to avoid such a risk. Sharma et al is derived a nonlinear model for the Tokamak devices based on the classical arguments of Hamiltonian mechanics, but then they derive a low-order model from it [16]. In this study, a general nonlinear model is presented and a nonlinear controller is designed for plasma magnetic control in a fusion reactor. The rest of the paper is organized as follows: Section 2 briefly discusses the modeling issues and then defines the control problem. Section 3 presents nonlinear control design for the system and also presents stability analysis of the controller. The last section highlights some concluding remarks.

2. Problem Statement

Dynamic model of a Tokamak is characterized by dynamic behaviors of externally excited poloidal field coils, induced eddy currents, plasma, and passive structure [17]. Before presenting a state-space model, some important assumption listed below should be addressed [2]:

- The plasma circuit system in a Tokamak is assumed to be axisymmetric. This assumption is discussed in detail in [18].
- The plasma current profile can be described by means of a finite number of global parameters [19].
- At the time scale of interest for current, position, and shape control, inertial effects can be neglected due to the low plasma mass density. Therefore, the plasma can be assumed to be in equilibrium at each time instant [2].
- Skin currents flowing on the plasma boundary can be neglected [20].

Considering these assumptions, circuit equations for the Tokamak model can be written as follows;

$$\begin{aligned}
&\frac{d\left(L_e I_e + M_{es} I_s + REI_e\right)}{dt} + \Omega_e I_e = V_e \\
&\frac{d\left(L_s I_s + M_{se} I_e\right)}{dt} + \Omega_s I_s = V_s \\
&\frac{d(m_e \dot{R})}{dt} = \frac{1}{2} I_e^T \frac{\partial L_e}{\partial R} I_e + I_s^T \frac{\partial M_{se}}{\partial R} I_e + \frac{1}{2} E I_e^2 \\
&\frac{d(m_e \dot{z})}{dt} = \frac{1}{2} I_e^T \frac{\partial L_e}{\partial z} I_e + I_s^T \frac{\partial M_{se}}{\partial z} I_e
\end{aligned} \tag{1}$$

where I_e is plasma elements current, I_s represents structure current, R, z depict plasma position at R- and z-axes of cylindrical coordinate system, L_e is plasma elements inductance, L_s stands for structure inductance, M_{se}, M_{es} are mutual inductances, Ω_e is plasma elements resistance, Ω_s represents structure resistance, E depicts internal energy coefficient, V_e stands for plasma elements voltage and V_s is poloidal coil voltage. System model given in (1) can be written in a more compact form as

$$\dot{x} = f(x,\theta) - g\left(x,\theta\right)u(t) \tag{2}$$

where the state vector is

$$x = \begin{bmatrix} I_e & I_s & \dot{r} \end{bmatrix}^T, \qquad r = \begin{bmatrix} R & z \end{bmatrix}^T \tag{3}$$

and the control input is

$$u = \begin{bmatrix} V_e & V_s & 0 \end{bmatrix} \tag{4}$$

and, finally, θ is the parameter vector containing constant parameters in (1). Note that

$$g(x,\theta) = I. \tag{5}$$

The generalized model given in (2) assumes that the system is linearly parameterizable. Such a parameterized form can be found in [16]. Control objective is to drive the components of the state vector, x, to a desired trajectory, x_d. To investigate the performance of the controller to be designed, a tracking error signal can be defined as

$$e = x_d - x. \tag{6}$$

It is assumed that the desired trajectory, x_d, is bounded with at least its second derivative.

3. Control Design

This section explains the design of a robust controller with three different versions to the control problem defined above. The main idea is to model the parameter uncertainties as a disturbance to the system and design the control signal so that the output is robust to these disturbances [21]. Investigating the corresponding open-loop error system dynamics yields

$$\dot{e} = \dot{x}_d - f(x,\theta) + u(t) \tag{7}$$

and, then, the robust controller formulation will be

$$u(t) = -\dot{x}_d - Ke + \hat{f} - V_R \tag{8}$$

where $\hat{f}$ is a function formed by the estimates of uncertain parameters vector, θ, and K is a positive definite control gain matrix. An estimation error signal can be defined as

$$\tilde{f} \triangleq f - \hat{f} \tag{9}$$

It is assumed that the estimation error function can be bounded by a known function, $\rho(\cdot)$, as

$$\left\|\tilde{f}(x,\theta)\right\| \le \rho\left(x,\hat{\theta}\right) \tag{10}$$

where $\hat{\theta}$ represents a vector containing the estimates of the elements of uncertain parameters vector, θ. In (8), V_R depicts the robust part of the controller to be designed. In this study, two versions for the robust part, that are, high frequency type and high gain type, are designed and presented in the followings. Then a third type of controller is introduced. Before presenting the design, expression for the error system dynamics at this step can be written as

$$\dot{e} = -Ke - \tilde{f} - V_R \tag{11}$$

3.1. High Frequency Type

Designing the robust part of the controller, VR, as

$$V_R = \frac{\rho^2 e}{\|\rho\|\|e\| + \varepsilon} \tag{12}$$

where $\varepsilon > 0$ is a small constant, yields a closed-loop error dynamics in the form of

$$\dot{e} = -Ke - \tilde{f} - \frac{\rho^2 e}{\|\rho\|\|e\| + \varepsilon}. \tag{13}$$

Following theorem and its proof can be used to analyze the effect of the composite control input signal given in (8) and (12) on the stability of the system given in (1).

Theorem 1: For the system given in (1), if the control input is designed as in (8) and (12), then the tracking error converges to a bound like

$$|e(t)| \le \sqrt{A + B\exp\{-2\lambda_{\min}\{K\}\}} \tag{14}$$

with an exponential envelope, where $\exp\{\cdot\}$ is the natural logarithm function, $\lambda_{\min}\{K\}$ is the smallest eigenvalue of K, and

$$A = \frac{\varepsilon}{\lambda_{\min}\{K\}}, \quad B = |e(0)|^2 - \frac{\varepsilon}{\lambda_{\min}\{K\}}. \tag{15}$$

Proof 1: To prove the theorem, a positive definite, scalar function can be used;

$$V = \frac{1}{2} e^T e. \tag{16}$$

Taking the time derivative of this function yields

$$\dot{V} = e^T \left\{ -Ke - \tilde{f} - \frac{\rho^2 e}{\|\rho\|\|e\| + \varepsilon} \right\} \tag{17}$$

which can be upper bounded as

$$\begin{aligned} \dot{V} &\leq -\lambda_{\min}\{K\}\|e\|^2 + \|\rho\|\|e\| - \frac{\rho^2 \|e\|}{\|\rho\|\|e\| + \varepsilon} \\ &\leq -\lambda_{\min}\{K\}\|e\|^2 + \varepsilon \left[\frac{\rho\|e\|}{\rho\|e\| + \varepsilon} \right] \leq -\lambda_{\min}\{K\}\|e\|^2 + \varepsilon. \end{aligned} \tag{18}$$

Considering (16) and (18) together, one can write

$$V(t) \leq V(0)\exp\{-\gamma t\} + \frac{\varepsilon}{\gamma}\left[1 - \exp\{-\gamma t\}\right] \tag{19}$$

where $V(0)$ is initial value of V, and

$$\gamma \triangleq 2\lambda_{\min}\{K\} \tag{20}$$

Then $e(t)$ can be upper bounded as in (14) and this completes the proof.

3.2. High Gain Type

For this type, the robust part of the controller is designed as

$$V_R = \frac{e\rho^2}{\varepsilon} \tag{21}$$

so that the corresponding closed-loop error system dynamics becomes

$$\dot{e} = -Ke - \frac{e\rho^2}{\varepsilon} - \tilde{f}. \tag{22}$$

Then the following theorem and its proof can be used to analyze the effect of the composite control input signal given in (8) and (21) on the stability of the system given in (1).

Theorem 2: The stability result given in Theorem 1 is also valid for the system if the control input is designed as in (8) and (21).

Proof 2: By using the same Lyapunov function given in (16) and taking again its time derivative, one gets

$$\dot{V} = -e^T Ke - \frac{\rho^2 e^T e}{\varepsilon} + e^T \tilde{f} \tag{23}$$

which can be upper bounded as

$$\dot{V} \leq -\lambda_{\min}\{K\}\|e\|^2 + \rho\|e\| - \frac{\rho^2\|e\|}{\varepsilon} \leq -\lambda_{\min}\{K\}\|e\|^2 + \rho\|e\|\left[1 - \frac{\rho\|e\|}{\varepsilon}\right]. \tag{24}$$

There exist two possible cases for the inequality given in (24).

Case 1: If $|\rho|\|e\| > \varepsilon$, then one can directly write

$$\dot{V} \leq -\lambda_{\min}\{K\}\|e\|^2. \tag{25}$$

Considering (16) and (25) together, it is inferred that $V(t)$ and its component, $e(t)$, goes to zero exponentially.

Case 2: If $|\rho|\|e\| < \varepsilon$, then

$$0 < \left(1 - \frac{\rho\|e\|}{\varepsilon}\right) \leq 1 \tag{26}$$

and

$$\dot{V} \leq -\lambda_{\min}\{K\}\|e\|^2 + \varepsilon \tag{27}$$

which gives same stability result in the high frequency type.

3.3. Smooth Robust Controller

Electrical drive systems used in the fusion reactors may suffer from chattering in system currents depending on the size of the reactor. Instant variations in the currents induce undesirable effects on fusion performance, since a linear amplifier cannot be used as the drive system due to the existence of high current requirement. Cooling such a linear amplifier would require some huge coolers with unreasonable dimensions. Instead, a more complex power electronics equipment are needed for drive system and these equipment generally suffer from rapid changes in the system currents created by chattering phenomena. Since this possibility is valid for all the systems with high current requirement, robust control researchers continuously seek more smooth robust schemes. For example, a useful topology is introduced in [22]. Tokamaks control systems may also require such topologies, again depending on the size of the reactor. This does not necessarily mean that high gain and high frequency type controllers introduced above are not useful and functional for Tokamaks. These controllers may still be successful for controlling plasma currents, shape, and position. Smooth robust topologies provide an alternative design approach to cope with some possible problems.

To investigate the smooth robust controller formulation for the system given in (1), let reconsider the error system dynamics given in (7), and define a new error variable, called filtered tracking error, as

$$\eta = \dot{e} + \alpha e \tag{28}$$

where $\alpha \in \mathbb{R}^+$ is a constant. Note that

$$e(s) = \frac{1}{s+\alpha}\eta(s) \tag{29}$$

which means that if $\eta\,(t)$ goes to zero, then $e(t)$ also goes to zero, and so does $\dot{e}(t)$, where s is Laplace variable. Investigating $\eta\,(t)$ dynamics yields

$$\begin{aligned} \dot{\eta} &= \ddot{e} + \alpha\dot{e} \\ &= \ddot{x}_d - \ddot{x} + \alpha\dot{e} = \ddot{x}_d + \alpha\dot{e} - \dot{f}(x,\theta) + \dot{u} - e + e. \end{aligned} \tag{30}$$

Let reconfigure $\eta\,(t)$ dynamics as

$$\dot{\eta} = N(\dot{x},\theta) - e + \dot{u} \tag{31}$$

where

$$N(\dot{x},\theta) = \ddot{x}_d + \alpha\dot{e} - \dot{f}(x,\theta) + e. \tag{32}$$

Now define

$$N_d \triangleq \ddot{x}_d + \frac{\partial f(x_d,\theta)}{\partial x_d}\dot{x}_d \tag{33}$$

in order to write

$$\dot{\eta} = \tilde{N} - e + \dot{u} + N_d, \quad \tilde{N} = N - N_d. \tag{34}$$

In (33), N_d represents the desired profile for N function. When the desired profile and its time derivatives are bounded, then N_d and its derivative are also bounded. Robust controller formulation assumes that $\tilde{N}$ and $\tilde{f}$ can be upper bounded by some functions, $\sigma(\cdot)$ and $\mu(\cdot)$, as

$$\begin{gathered} \|\tilde{N}\| \le \sigma(\|\varphi\|)\|\varphi\|, \quad \varphi \triangleq [e \quad \eta]^T \\ \|\tilde{f}(x,\theta)\| \le \mu(x,\hat{\theta}) \end{gathered} \tag{35}$$

where $\tilde{f} = f - \hat{f}$, $\hat{f}$ is the estimation of f function, and $\hat{\theta}$ is the estimation of unknown parameters vector. Now one can design $\dot{u}$ as

$$\dot{u} = -\left[(K_s + I)\eta + \beta \operatorname{sgn}(e)\right] \tag{36}$$

where β and K_s are some positive control gain matrices and I is identity matrix. Then the closed-loop error system dynamics becomes

$$\dot{\eta} = -(K_s + I)\eta - \beta \operatorname{sgn}(e) + N_d - e + \tilde{N}. \tag{37}$$

The implementable form of the actual control input signal, u, will be given at the end of the following stability analysis. The following theorem and its proof can be used to show the stability of the overall system.

Theorem 3: For the system given in (1), the control input signal defined by (37) ensures that the tracking error signal converges to as time goes to infinity, if the minimum eigenvalues of the control gain matrices are selected to satisfy

$$\lambda_{\min}\{\beta\} > \|N_d(t)\| + \frac{1}{\alpha}\|\dot{N}_d\|, \quad \text{and} \quad \lambda_{\min}\{K_s\} \ge \frac{1}{4}\sigma^2(\|\varphi\|). \tag{38}$$

The designed control input is also guarantees that all signals in the closed-loop system are bounded.

Proof 3: The proof is omitted due to lack of space.

4. Conclusion

Three different versions of a nonlinear robust controller for plasma current, shape, and position control in a nuclear fusion reactor were presented. The first type, high frequency controller, drives the plasma current, shape, and position to a desired trajectory with an adjustable error, while the second type, high gain controller, may drive the tracking error signal to zero by condition, or, at least to an adjustable bound. If the worst case of these controllers, that is, the tracking error signal converges to an adjustable bound, is occurred, this may lead chattering in poloidal coils voltage and plasma elements voltage, and in currents created by these voltages. But, still the control objective is achieved. The third controller provides an alternative robust control strategy for the reactor and aims to reduce chattering. All three controllers provide global stability result, that is, global confinement of plasma is achieved.

References

[1] Ariola M and Pronti A 2008 *Proc. 47th IEEE Conf. on Decision and Control* 3119-24
[2] Ambrosino G and Albanese R 2005 *IEEE Control Systems Magazine* **5** 76-92
[3] www.efda.org
[4] Belyakov V A and Kavin A A 2003 *Proc. Int. Conf. on Physics and Control* 1019-24
[5] Garrido A J, Garrido I, Barambones O, Alkorta P and Maseda F J 2008 *Proc. Int. Conf. on Control* 2429-32
[6] Mitrishkin Y V, Dokuka V N, Khayrutdinov R R and Kadurin A V 2006 *Proc. 45th IEEE Conf. on Control and Decision* 2207-12
[7] Moreau D, Crisanti F, Litaudon X, Mazon D, De Vries P, Felton R, Joffrin E, Laborde L, Lennholm M, Murari A, Ridolfini V P, Riva M, Tala T, Tresset G, Zabeo L and Zastrow K D 2003 *Nuclear Fusion* **43** 870-82
[8] Laborde L, Mazon D, Moreasu D, Murari A, Felton R, Zabeo L, Albanese R, Ariola M, Bucalossi J, Crisanti F, De Baar M, De Tommasi G, De Vries P, Joffrin E, Lennholm M,

Litaudon X, Pironti A, Tala T and Tuccillo A 2005 *Plasma Physics and Controlled Fusion* **47** 155-63
[9] Moreau D, Mazon D, Ariola M, De Tommasi G, Laborde L, Piccolo F, Sartori F, Tala T, Zabeo L, Boboc A, Bouvier E, Brix M, Brzozowski J, Challis C D, Cocilovo V, Cordoliani V, Crisanti F, De La Luna E, Felton R, Hawkes N, King R, Litaudon X, Loarer T, Mailloux J, Mayoral M, Nunes I, Surrey E and Zimmerman O 2008 *Nuclear Fusion* **49** 1-38
[10] Barana O, Mazon D, Laborde L and Turco F 2007 *Plasma Physics and Controlled Fusion* **49** 947-67
[11] Suzuki T, Isayama A, Ide S, Fujita T, Oikawa T, Sakata S, Sueoka M, Hosoyama H and Seki M 2005 *Proc. 16th Topical Conf. on Radio Frequency Power in Plasmas* 279-286
[12] Xu C, Ou Y, Dalessio J, Schuster E, Luce T C, Ferron J R, Walker M L and Humphreys D A 2010 *IEEE Transaction on Plasma Science* **38** 163-73
[13] Hyatt A W, Ferron J R, Humphreys D A, Chamberlain F R, Johnson R D, Penaflor B G, Piglowski D A, Scoville J T and Walker M L 2010 *IEEE Transaction on Plasma Science* **38** 434-40
[14] Shpanin L M, Jones G R, Spencer J W and Djakow B E 2008 *IEEE Transaction on Plasma Science* **36** 2795-800
[15] Cenedese A and Sartori F 2003 *Proc.42nd IEEE Conf. on Control and Decision*, 4628-33
[16] Sharma A S, Limebeer D J N, Jaimoukha I M and Lister J B 2005 *IEEE Trans. on Control Systems Technology* **13** 356-69
[17] Garrido I, Garrido A, Barambones O and Alkorta P 2007 *Proc. 6th WSEAS Int. Conf. System Science and Simulation in Engineering* 109-14
[18] Ward D J and Jardin S C 1992 *Nuclear Fusion* **32** 973-93
[19] Albanese R, Fresa R, Rubinacci G and Villone F 2000 *IEEE Transactions on Magnetics* **36**, 1804-07
[20] Albanese R and Villone F 1998 *Nuclear Fusion* **38** 723-38
[21] Zergeroglu E, Kandemir I, Seker M and Eroglu E J. Fac. Eng. Arch. Gazi Univ. 2006 **21** 587-602
[22] Xian B, De Queiroz M S and Dawson D M 2004 *Lecture Notes in Control and Information Sciences:Optimal Control, Stabilization, and Nonsmooth Analysis* Springer Verlag, 251-66

IOP Conf. Series: Materials Science and Engineering **261** (2017) 012016 doi:10.1088/1757-899X/261/1/012016

A Cognitive Computing Approach for Classification of Complaints in the Insurance Industry

J Forster[1, 2,*] and B Entrup[2]

[1]Department for Computer Sciences and Mathematics, University of Applied Sciences Munich, Lothtraße 64, 80335 Munich, Germany
[2]Global Business Services, IBM Germany, IBM-Allee 1, 71139 Ehningen, Germany

*jforster@hm.edu

Abstract. In this paper we present and evaluate a cognitive computing approach for classification of dissatisfaction and four complaint specific complaint classes in correspondence documents between insurance clients and an insurance company. A cognitive computing approach includes the combination classical natural language processing methods, machine learning algorithms and the evaluation of hypothesis. The approach combines a MaxEnt machine learning algorithm with language modelling, tf-idf and sentiment analytics to create a multi-label text classification model. The result is trained and tested with a set of 2500 original insurance communication documents written in German, which have been manually annotated by the partnering insurance company. With a F1-Score of 0.9, a reliable text classification component has been implemented and evaluated. A final outlook towards a cognitive computing insurance assistant is given in the end.

1. Introduction

Every day we create 2.5 quintillion bytes of data in all kinds of industries and according to the widely quoted 80 % rule, most of this enterprise data is generated and stored in an unstructured form such as text and images [1]. This applies in particular to the insurance industry, where medical assessments, claim reports, bills, blanks and letters arrive every second. The process automation quota is growing steadily, for example for property/casualty insurers from 11.8 % in 2013 to 15.3 % in 2014 [2]. However, this automation relies on structured or at least semi-structured documents and data. Yet, companies are aware of the importance of their internal and external unstructured data, while only 25 % of unstructured data is used according to Forrester [1] as one can see in figure 1.

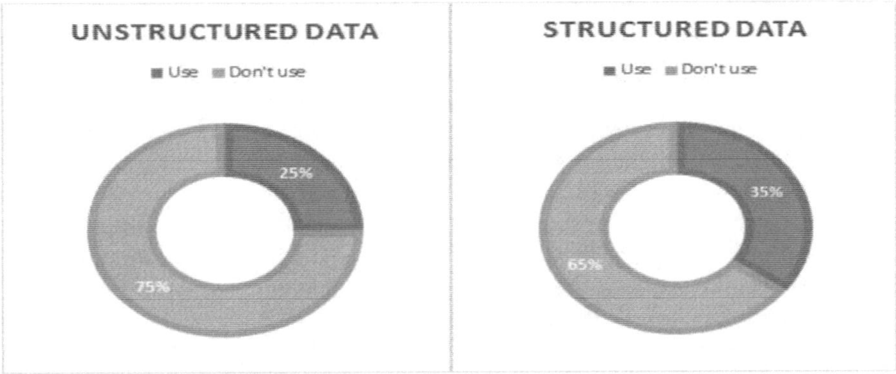

Figure 1. Percentage of currently used data for BI in insurance companies [1]

2. Problem Definition

A large part of the growing challenge regarding unstructured data in the insurance industry is related to the change of communication. With the ease of writing an email instead of a letter as well as the rise of mostly internet-dependent direct insurance companies comes an increase of digital correspondence. These have to be handled by the input management systems of these companies. While semi-structured data, such as bills and forms can be handled by rule-based classification systems, cover letters are completely unstructured. However, valuable information can be extracted from it and be used for a more efficient routing in an insurance company.

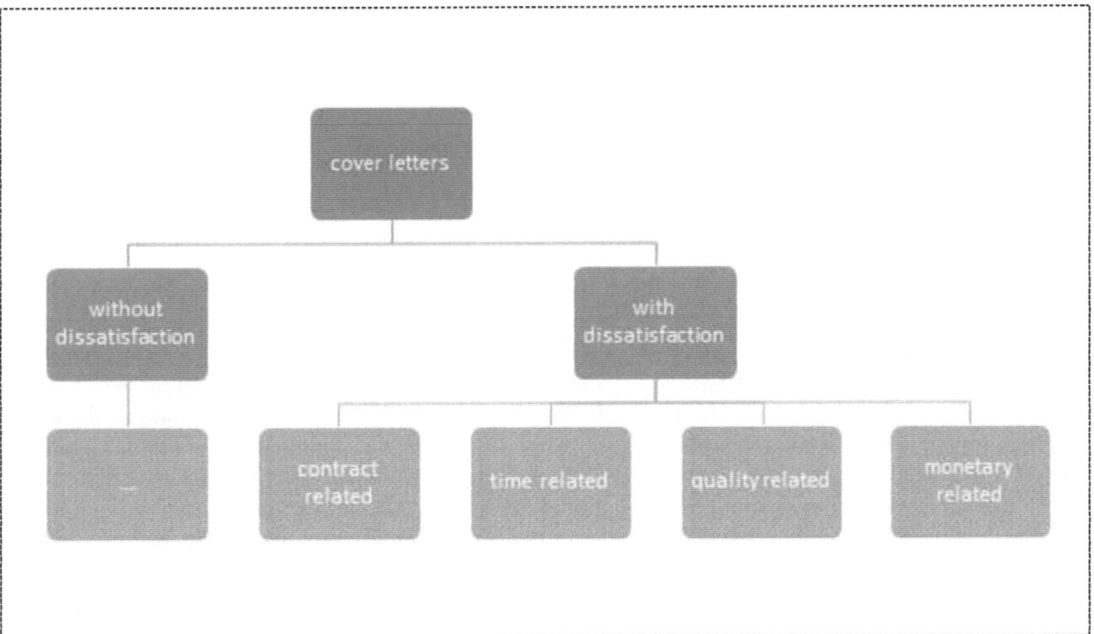

Figure 2. Classification hierarchy of complaints

In this paper, a machine learning approach for handling unstructured insurance cover letters is evaluated on the example of complaint letters. As one can see in figure 2, the classification of an complaint has two stages. In the first stage, the letters without dissatisfaction have to be separated from the letters, which could contain a complaint. In the second stage, the reason of the complaint has to be identified to ensure better case handling. In the case of assigning a dissatisfaction type, we have

4 classes: 'contract related', 'time related', 'quality related', and 'monetary related'. There are two general strategies for solving such a multi-class classification problem: one-vs-all (OVA) and one-vs-one (OVO). While OVA requires one binary classifier classifying each class against all other classes (4 in our case), OVO requires one classifier classifying each combination of classes (4*(4 – 1) / 2 = 6 in our case). We select OVA as our strategy, due to its conceptual simplicity and because it has been shown to be as accurate as any other approach if finely-tuned regularized classifiers are used [3].

2.1. Limitations of Current Solution

Currently deployed input management systems rely heavily on static rules. These can identify standardized formulas or even react to specific keywords in continuous text. Multiple problems arise when dealing with cover letters from customers:

- Terminology: Insurance customers in general have highly varying backgrounds and therefore use different words and syntax, especially within complaints. In the test set used in this paper and described in more detail later, lawyers for example complain in a more complex and extensive way about an incompliant process than a regular insurance holder who is unhappy with the quality of service. Furthermore, brokers who complain on behalf of their customers tend to use more professional insurance terms than regular customers.
- Context: Important or subliminal information is often hidden in the greater context of a cover letter. Especially dissatisfaction can be expressed in a subliminal form. The request for sending a contract cancelation confirmation can be written as "Please send me a written confirmation of the contract termination" or "This is now the 3rd time I have to remind you to send me the confirmation of the contract termination". Both have the same intention but with a different sentiment.
- Equivalence: In domains where there is a special terminology used by professionals but also laypersons, words are meant differently as used by the different groups. For example, in the insurance context of this paper an insurance policy, insurance contract and an insurance note have different meanings for the professional clerks, but are treated as synonyms by the clients.

Still, all these ambivalent and unstructured messages have to be processed by input management systems, which are not able to satisfactorily distinguish between two similar sentences as shown above. Until now only human brains have the cognitive capability to handle unstructured data.

2.2. Cognitive Computing

This paper evaluates a solution to overcome the above mentioned problems with unstructured text. Kelly et al. introduced the concept of cognitive computing [4]. The philosophy behind cognitive computing is the combination of various technologies to 'understand' unstructured data such as text, images, sound and video like a human brain is capable of. Real cognitive computing consists of three columns as depicted in figure 3. First of all the unstructured data has to be made analysable for machines. Image recognition technologies are used while natural language processing makes text handling possible. The next column represents the human capability of learning with various machine learning techniques such as neuronal networks, deep learning algorithms or decision trees.

The third and final column is the most challenging to achieve: the human capability of creating hypothesis and reason. Diverse statistical algorithms are designed, trained and tested for each single use case.

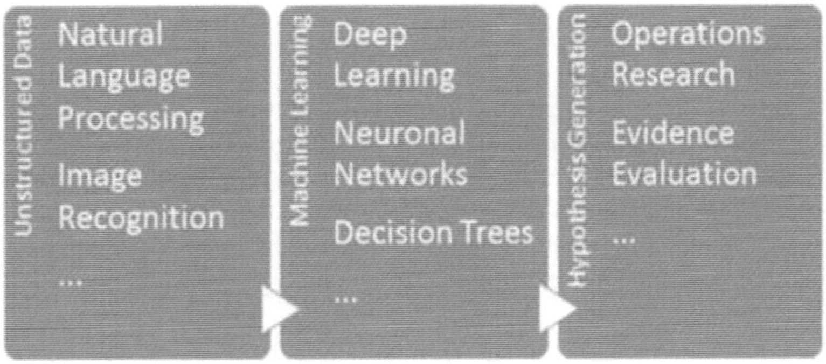

Figure 3. The three columns of Cognitive Computing

3. Related Work

The work presented in this article is related to the research areas of machine learning, natural language processing and document classification. These areas and the related work are used together to create a cognitive computing approach for document classification in a real life insurance environment and reduce the problems described in chapter two occurring in this environment.

3.1. Document Classification

The automatic classification of text has a long history in information technology research [5]. Text classification tasks can be divided into single-label (one category is assigned to each text) and multi-label (several categories can be assigned) cases. While there are also dynamic and static rule based classification systems, we want to concentrate on adaptive machine learning classification systems. Kailer et al. describe and evaluates different state of the art classification models but also highlight the ambivalence depending on each practical use case and the underlying data set [6]. The in this research used data set of insurance related correspondence has not yet been assessed. The creation of a new cognitive computing based text classification approach is the general purpose of this research.

3.2. Machine Learning

In the last decade, binary machine learning algorithms such as decision trees [7], probabilistic classifiers such as Naive Bayes [8], support vector machines [9], and maximum entropy (MaxEnt) classifiers have been applied successfully to the task of document classification [10]. In the case of multiclass classification, these are combined either according to a one-vs-one or one-vs-all scheme.
Rifkin et al. show that the latter given finely-tuned binary classifiers is at least as accurate as more complex approaches. Recently, Kim has achieved state-of-the-art results on several text classification benchmark tasks using convolutional neural networks [11]. The integration of machine learning reduces the 'context' problem.

3.3. Sentiment Analysis

The field of Sentiment Analysis deals with extracting opinion, mood and attitude from unstructured data. This task can be performed on different granularity levels. When whole documents are the unit of analysis, this is carried out as classification where the labels are so-called polarities e.g. positive or negative [12].

A popular approach is the use of a lexicon containing words annotated with these polarities. A more challenging problem arises when one tries to grade the strength of the opinion on a numeric scale (rating inference problem) [13].

We use simple heuristics to tackle this problem. They make use of the lexicon to infer the rating for a sentence from its syntax. These ratings are then combined to form a document rating. The usage of sentiment analysis also reduces the 'context' problem.

3.4. Language Modelling

Latent dirichlet allocation (LDA) or topic modelling is a generative process that allows the discovery of latent themes that underlie a set of documents [14]. Topic modelling algorithms have been adapted and used in a wide array of applications, such as to find patterns in genetic data, images, and social networks, but are particularly popular in information retrieval, classification, and corpus exploration tasks as they allow the organization of a large collection of unstructured texts according to its main topics. Topic models have been extended, e.g. to consider the change of topics over time [14], hierarchies of topics [15], as well as metadata such as authors [16] or inter-document links [17]. Language Modelling can lessen the 'terminology' issue.

3.5. Tf-idf

Term frequency-inverse document frequency (tf-idf) is a numerical method to compute the relevance of a term in a document given a corpus of documents. For a word we use its raw frequency (tf_t) as well as the so called "inverse document frequency"(idf_t), which was introduced 1972 .The latter measures the significance of a term t in a corpus by comparing the document frequency df_t , which is the number of documents that contain t, and the total number of documents N.Combined, these notions yield a composite value for each term in a document:

$$tf - idf_t = tf_t \times idf_t = tf_t \times \log \frac{N}{df_t} \tag{1}$$

See [18] for more information. Implementing tf-idf in our architecture tackles the 'equivalent' problem.

4. Cognitive Computing Approach

To evaluate the usefulness and feasibility of a Cognitive Computing document classifier we used the above mentioned technologies in combination to fulfil two of the three steps towards a cognitive computing system.

The implanted UIMA-based dissatisfaction classifier classifies between dissatisfaction and non-dissatisfaction as well as between different dissatisfaction types using a binary classifier for each. It can be trained and tested on two disjoint sets or can be evaluated using cross-validation on a single set of documents. It employs various lexical and semantic features as well as topic models and word features for classification.

4.1. UIMA

The UIMA (Unstructured Information Managment Architecture) standard has been developed by IBM and is the defacto standard for analysing unstructured data. The basic architecture is built as an UIMA pipeline as depicted in figure 4.

Figure 4. Classical UIMA pipeline

The text elements or documents are handled from left to right, with basic natural language capabilities in the beginning to more sophisticated and use case specific modules at the end [19].

IOP Conf. Series: Materials Science and Engineering **261** (2017) 012016 doi:10.1088/1757-899X/261/1/012016

4.2. Orchestration
On top of the UIMA architecture we use the ClearTK toolkit, developed at the University of Colorado. It is a natural language processing framework which offers interfaces to different machine learning libraries [20].

5. Experiment
In the following we are training and evaluating a MaxEnt algorithm on documents manually annotated by domain experts. This documents consist of emails, letters and faxes which have been actually sent by insurance customer to the insurance company cooperating in our research project.

5.1. Dataset
For training and evaluation, domain experts manually annotated a training set of 2,500 documents and a test set of 1,500 documents (This almost evenly split set was a requirement of the involved insurance company). The training set contains 1,500 texts with complaints and 1,000 non-complaints. Both sets contain documents that have been assigned one or more labels indicating the type of complaint expressed in the text. Each text can contain up to four different classes since one text can express several complaints or one complaint can belong to several of the possible classes.

5.2. Learning and Testing Procedure
Before starting the actual feature extraction and training, a number of pre-processing steps were applied. First, the texts were cleaned: header and footers of e-mails and letters were removed, and the texts were anonymized. Furthermore, the texts were tokenized, lemmatized, and part-of-speech (POS) tagging was applied. Then the actual features were extracted. Besides a bag-or-words approach using all lemmas, synonyms were inserted, bi-grams extracted, and the lemmas with the highest tf-idf and log-likelihood values were calculated and sorted by their POS. Besides these features, each text was assigned a set of topic models and the sentiment model described above was used to calculate first a value for each word and combine these values for an overall value for each text. Some other features based on pattern matching, e.g., the use of caps-lock and the number of exclamations marks, were also used but will not be discussed and will not be evaluated in detail here. In total, five different classes were investigated: does the text contain a complaint? If so, does it concern the quality or the time of the treatment or processing, does it concern the contract conditions, or does it concern the payment? The latter four are or core type of complaints and each text can be assigned more than one of these classes.

5.3. Evaluation
Although the project described here included several sub-classes that were successfully implemented, only the binary classification complaint/no-complaint classification can be considered and evaluated here. The overall final results that were obtained are very satisfying. Complaints could be identified with a precision of 0.84 and a recall of 0.97, resulting in an f1-score of 0.90. To evaluate the impact of the chosen features on the final results, an ablation study is performed, i.e., each feature is removed one after another, the model is trained and evaluated again, and the difference, i.e., the impact, of leaving out a single feature is measured by the difference in f1-score.

Table 1. Ablation study

Feature	Precision	Recall	F1	Differences
Baseline	0.84	0.97	0.90	0.00
Bag-of-words	0.59	0.87	0.70	-0.20
Synonyms	0.84	0.97	0.90	0.00

Tf/idf	0.77	0.96	0.85	-0.05
Topic models	0.82	0.96	0.88	-0.02
Sentiment	0.83	0.97	0.89	-0.01
Regex	0.70	0.94	0.80	-0.10
Bi-grams	0.82	0.95	0.88	-0.12

As can be seen from table 1, most of the presented features contribute at least partially to the overall performance. The bag-of-words approach has the highest impact if left out. The remaining features only increase the f1-score by 0.2 points. Additional information is also added by the tf-idf based features, i.e., the highest scoring words of the POS noun, verb, and adjective. Moreover, a large impact can be found when leaving out the regex-based features. These patterns are based on experience and a looking for keywords that seem to be indicating a possible complaint. The topic models and the sentiment score based on a sentiment dictionary especially tuned to the insurance sector contribute to the final model but to a lower degree. The domain specific tuning has to be done because standardized sentiment analysis tend to give an insurance related letter a negative rating as they are often related to damage, losses, illness etc. However this is the daily business of the insurance industry and cannot be treated as a negative sentiment.

6. Conclusion

The promising first results depicted in table 1 lead now to an ongoing research activity within the insurance company as well with and integration into the current document process flow. Furthermore the cognitive computing approach will be used for enhancing the user experience at the workplace of each insurance clerk handling client cases. A first step has been made by implementing a text summary tool based on the text rank algorithm [21].This leads to a cognitive advisor tool for insurance clerks in the near future as the proven cognitive computing approach can now be adopted for more detailed domain specific document classes. Additionally the domain specific sentiment analysis will be evaluated and enhanced in a separate research approach.

References

[1] B. Evelson, L. Owens und B. Hopkins, Market Overview: Text Analytics, 2015.
[2] M. Hundt, Versicherer-Kommunikation: Papier dominiert weiter, 2015.
[3] R. Rifkin und A. Klautau, In defense of one-vs-all classification, The Journal of Machine Learning Research, Bd. 5, pp. 101-141, 2004.
[4] J. Kelly III und S. Hamm, Smart Machines: IBM's Watson and the Era of Cognitive Computing, Columbia University Press, 2013.
[5] H. Borko und M. Bernick, Automatic document classification, Journal of the ACM (JACM), Bd. 10, pp. 151-162, 1963.
[6] D. Kailer, P. Mandl und A. Schill, Grouping Product Aspects from Short Texts Using Multiple Classifiers, in Web Information Systems Engineering – WISE 2015, Springer International Publishing, 2015, pp. 1-15.
[7] W. W. Cohen und H. Hirsh, Joins that Generalize : Text Classification Using WHIRL, KDD-98 Proceedings, pp. 169-173, 1998.
[8] D. Koller und M. Sahami, Hierarchically Classifying Documents Using Very Few Words, Proceedings of the Fourteenth International Conference on Machine Learning, pp. 170-178, 1997.
[9] T Joachims, Text Categorization with Suport Vector Machines: Learning with Many Relevant Features, Proceedings of the 10th European Conference on Machine Learning ECML '98, pp. 137-142, 1998.
[10] H. L. Chieu und H. T. Ng, A maximum entropy approach to information extraction from semi-structured and free text, AAAI/IAAI, Bd. 2002, pp. 786-791, 2002.

[11] Y. Kim, „Convolutional Neural Networks for Sentence Classification," 2014.
[12] Liu, Sentiment Analysis Mining Opinions, Sentiments, and Emotions, Cambridge University Press, 2015.
[13] B. Pang und L. Lee, Seeing Stars: Exploiting Class Relationships for Sentiment Categorization with Respect to Rating Scales, in Proceedings of the 43rd Annual Meeting on Association for Computational Linguistics, Stroudsburg, 2005.
[14] D. M. Blei, A. Y. Ng und M. I. Jordan, Latent Dirichlet Allocation, Journal of Machine Learning Research, Bd. 3, pp. 993-1022, 2003.
[15] D. Blei und J. Lafferty, Dynamic topic models, Proceedings of the 23rd international conference {\ldots}, pp. 113-120, 2006.
[16] M. Rosen-Zvi, T. Griffiths, M. Steyvers und P. Smyth, The author-topic model for authors and documents, Proceedings of the 20th conference on Uncertainty in artificial intelligence, pp. 487-494, 2004.
[17] J. Chang und D. M. Blei, Hierarchical relational models for document networks, The Annals of Applied Statistics, Bd. 4, pp. 124-150, 2010.
[18] C. D. Manning, P. Raghavan und H. Schütze, Scoring, term weighting and the vector space model, Introduction to Information Retrieval, Bd. 100, pp. 2-4, 2008.
[19] D. Ferrucci und A. Lally, UIMA: an architectural approach to unstructured information processing in the corporate research environment, Natural Language Engineering, Bd. 10, pp. 327-348, 2004.
[20] P. V. Ogren, P. G. Wetzler und S. Bethard, ClearTK: A UIMA toolkit for statistical natural language processing, Towards Enhanced Interoperability for Large HLT Systems: UIMA for NLP, Bd. 32, 2008.
[21] R. Mihalcea und P. Tarau, TextRank: Bringing order into texts, 2004.

IOP Conf. Series: Materials Science and Engineering **261** (2017) 012017 doi:10.1088/1757-899X/261/1/012017

A Novel Recommendation System to Match College Events and Groups to Students

K Qazanfari[1, *], A Youssef[1], K Keane[2] and J Nelson[2]

[1] Department of computer science, George Washington University, Washington, DC, USA 20052
[2]Promantus Inc. Cary, NC, USA 27511

*Kazemmit@gwu.edu

Abstract. With the recent increase in data online, discovering meaningful opportunities can be time-consuming and complicated for many individuals. To overcome this data overload challenge, we present a novel text-content-based recommender system as a valuable tool to predict user interests. To that end, we develop a specific procedure to create user models and item feature-vectors, where items are described in free text. The user model is generated by soliciting from a user a few keywords and expanding those keywords into a list of weighted near-synonyms. The item feature-vectors are generated from the textual descriptions of the items, using *modified tf-idf* values of the users' keywords and their near-synonyms. Once the users are modeled and the items are abstracted into feature vectors, the system returns the maximum-similarity items as recommendations to that user. Our experimental evaluation shows that our method of creating the user models and item feature-vectors resulted in higher precision and accuracy in comparison to well-known feature-vector-generating methods like Glove and Word2Vec. It also shows that stemming and the use of a modified version of tf-idf increase the accuracy and precision by 2% and 3%, respectively, compared to non-stemming and the standard tf-idf definition. Moreover, the evaluation results show that updating the user model from usage histories improves the precision and accuracy of the system. This recommender system has been developed as part of the Agnes application, which runs on iOS and Android platforms and is accessible through the Agnes website.

1. Introduction

With a world of exponential growth of online information, finding useful data can be surprisingly difficult [1]: there are too many options for users to assess, sort through, and find the ones most relevant to their interests. To remedy the information overload, recommender systems could be a valuable tool for online users [1].

Recommender Systems or Recommendation Systems (RSs) are a subclass of information filtering systems that collect information on the preferences of users for a set of items such as movies, songs, books [2, 3]. RS might use different user information such as age and nationality, or social information such as followers, followed, tweets, and posts [2]. RSs compare the user's information to available items against reference characteristics in order to find recommendations [1]. Generally, a recommendation system compares the user profile to some reference attributes to predict the 'rating' that a user would give to an item that has not been seen by the user [1].

Generally, the recommender systems are classified into the following three categories:

- Collaborative recommendation systems: These approaches recommend the items based on the similarity of tastes and preferences between users using a utility matrix [4, 5].
- Content-based recommendation systems: These approaches recommend the items based on the description of the item and a profile of the user's preference and history [6-10].
- Hybrid recommendation systems: These approaches combine collaborative and content-based methods to recommend some items to the users [11, 12].

Since the content-based systems are designed mostly to recommend text-based items [6], the rest of this subsection will focus on reviewing of some text-content-based recommender systems. Recently, considerable research has been done in the field of text-based recommendation systems. Gu et al. [7] proposed a method to measure the similarity between two pieces of text using cosine similarity of the two bags of words, where each word is weighted by its tf-idf, and they applied this method on LinkedIn job data. A similar method has been proposed by Philip et al. [8] but for digital libraries. Also, Lak et al. [9] proposed an article recommendation system that uses tf-idf and word2vec for converting each article to a feature-vector. Some more comprehensive notion of document relevance than bags-of-words are proposed in [13, 14]; they extract a wide variety of content-based features to characterize non-linguistic aspects of the audio such as speaker, language, gender, and environment. In [13], authors use Glove for feature extraction and using them into CNN.

Although hybrid recommendation systems might provide higher performance, the collaborative RS component needs sufficient rated items to create the utility matrix. However, since often there is not enough rated items in exclusive communities for creating the utility matrix, this paper proposes a novel text-content-based recommendation systems for recommending some items (e.g. college events) to users (e.g. students). By considering the unstructured text content of the items, we convert the content of these items to some meaningful feature-vectors by computing a modified version of tf-idf on some specific pre-processed words. Finally, we find the similarity between feature-vectors using cosine measurement. The details of the proposed method are presented in the next section.

The rest of this paper is organized as follows: The proposed recommendation system will be explained in Section 2. In Section 3, experiments will be presented and analyzed. The paper concludes in Section 4 with a discussion of the results achieved and some suggestions of lines of future work in the field.

2. Proposed recommendation system

In this section, we propose a novel content-based recommendation system to increase the user engagement in exclusive communities. In the content-based recommender systems, the recommendation decisions are made based on the item content and the user content. In our method, items are activities that take place at a specific time and specific place, and described in free text, like the following item describing an event in a college community:

Event Title, e.g., "Smithsonian American Art Museum Highlights Tour"

Event Description, e.g., "This highlights tour of the museum's collection may include highlights of the temporary exhibitions. Location: F Street Lobby."

Event Tags, e.g., Art, Museum

Also, the user content includes a few keywords that are solicited from a user and that represent his/her interests, hobbies, and needs such as Painting, Cooking, Fitness and Traveling. The following gives the main processes of our recommendation system:

A. Creating the initial user model for each new user based on keywords from the user.
B. Updating the user model based on user history (negative and positive items).
C. Recommending some new matched items to the user based on his/her model.

Each of the above processes will be explained and illustrated in the following sections.

2.1. Creating the initial user model

Based on the targeted community, each user has some interests, and the intensity of interests should be quantified. For example, if a user is very interested in "sport", we model it as {sport: 5}, or if a user is not interested in "sport", we model it as {sport: -2}, so these numbers reflect the intensity of user interest.

In this section, we introduce the process of creating the initial user model; an array of {word: weight} pairs. This array will be created based on the keywords provided by the user.

The user model includes two types of word: the keywords that are provided by the user directly; and words that are the *near-synonyms* of those keywords. The near-synonyms of a keyword are the words that are synonymous or similar (in meaning) to the keyword. The number of near-synonyms of a keyword is a parameter (S) that is set before creating the initial user model.

Regarding the keyword weight, we assign initially a constant W_{max}. The weight of a near-synonym of a keyword ($W_{\text{near-synonym}}$), is a number between 0 (exclusive) and 1 (inclusive), where higher weights signify more similarity. These near-synonym weights are computed by the near-synonym generator such as Glove. For example, if a user provides "sport" and "technology" as keywords, and we take $S = 5$ and $W_{max} = 2$, then the process of creating the initial user model is:

1) Collecting S=5 near-synonyms of the "sport" and "technology" keywords as shown in Table 1:

Table 1. Near-synonyms of the "sport" and "technology" keywords.

near-synonyms of "sport":	athletics	football	rowing	racing	wrestling
$\boldsymbol{W}_{\textbf{near-synonym}}$:	1.0	1.0	0.9	0.9	0.8
near-synonyms of "technology":	engineering	IT	application	business	technological
$\boldsymbol{W}_{\textbf{near-synonym}}$:	0.8	0.8	0.6	0.5	0.5

2) Creating the initial user model by computing the union of all keywords and their near-synonyms. If there is a duplicated word, use the highest weight among all duplicates as shown in Table 2.

Table 2. A sample initial model based on "sport" and "technology" keywords.

word:	sport	athletics	Football	Rowing	Racing	Wrestling	technology	engineering	IT	application	business	technological
Weight:	2.0	1.0	1.0	0.9	0.9	0.8	2.0	0.8	0.8	0.6	0.5	0.5

2.2. *Refining the user model and finding top recommended items for users*

In this section, we will explain how to refine the user model and find the top N matched items for each user. From the viewpoint of having useful information, we suppose that the content of each item has different textual parts; we call each part a *data field*. For example, an event item in a college community might have three different data fields as Event Title, Event Description and Event Tags, and each data field has a different amount of useful information, which is called the *significance weight* of the data field. The significance weights are parameters in our system, set at configuration time of the RS.

To find the top N matched items, the following steps are performed:

1) Refine the user model into a model $User_{RM}$ as follows:
    a. Calculate the lemmatized and/or stemmed versions of the initial/updated user model and call the resulting model $Model_L$ and/or $Model_S$
    b. For each keyword or near-synonym, remove the last letter of the word if the word ends with 'e', 'y' or 'i' based on the following conditions:
        i. After lemmatizing, if a word ends with 'e' or 'y', remove the last letter

ii. After stemming, if a word ends with 'i', remove the last letter

c. We allow lemmas and stems of length at least 3 letters. Therefore, after performing steps 1.a and 1.b, remove all words of length 2 or less from $Model_L$ and $Model_S$

d. If we only use the lemmatized version of the user model, set $User_{RM} = Model_L$

e. If we only use the stemmed version of the user model, set $User_{RM} = Model_S$

f. If we use both lemmatized and stemmed versions, set $User_{RM}$ to be the union of $Model_L$ and $Model_S$

g. In all the above cases, if we have a duplicated word, we should keep only one copy of it and use the highest weight among all duplicates as the word weight

2) Let U_v be the weight vector in e $User_{RM}$

3) For each item i, calculate the similarity W_i between the item feature-vector S_i and user weight vector U_v as follows:

a. For each data field j with significance weight $W_{df(j)}$:

i. $S_{df(j)}$:= the tf-idf for the jth data field of the item for the words that we have in the $User_{RM}$;

b. Calculate the item feature-vector S_i by weighted summation over all data fields:

$$S_i = \sum_{j=1}^{n=number\ of\ data\ fields} S_{df(j)} * W_{df(j)} \quad (1)$$

c. Calculate W_i as the similarity between S_i and U_v; (similarity measures will be addressed later)

4) Sort all items based on calculated W's in decreasing order.

5) Select top N items.

To make the above processes clear, we illustrate them with an example: Consider the initial user model as shown in Table 2, and using stemming for Step 1.a. Also, suppose that the target community has items with three data fields.

In this example, we use the stemmed version of the user model for creating $User_{RM}$. Based on the mentioned user model, $Model_S$ will be changed as shown in Table 3. In this table, for each step, the cells which were affected by this step are colored in gray.

Table 3. Refining the user model based on the initial model of Table 2.

User model words:	**Step 1.a.**	**Step 1.b.**	**Step 1.c.**
sport	sport	sport	sport
athletics	athlet	athlet	athlet
football	footbal	footbal	footbal
rowing	row	row	row
racing	race	race	rac
wrestling	wrestl	wrestl	wrestl
technology	technolog	technolog	technolo
engineering	engin	engin	engin
IT	IT	IT	
application	applic	applic	applic
business	busi	Bus	bus
technological	technolog	technolog	technolo

Step 1.g.	sport	athlet	fotbal	row	rac	wrestl	technolog	engin		applic	bus	
Weight	2.0	1.0	1.0	0.9	0.9	0.8	2.0	0.8	0.8	0.6	0.5	0.5

Finally, Table 4 shows the refined model $User_{RM}$.

Table 4. Refined model on the initial model of Table 2.

$User_{RM}$ **words:**	sport	athlet	footbal	row	race	wrestl	technolog	engin	applic	bus
$User_{RM}$ **words vector** (U_v):	2.0	1.0	1.0	0.9	0.9	0.8	2.0	0.8	0.6	0.5

Now, suppose we have 3 different college events as items. The first one is about honoring a football team, the second item is about course registration, and the last one is about using technology in health. The calculated tf-idf of each mentioned item for $User_{RM}$ words are as mentioned in Table 5.

Table 5. Calculated tf-idf for $User_{RM}$ words for three items.

$User_{RM}$ **words:**		sport	athlet	footbal	row	race	wrestl	technolog	engin	applic	bus
	$S_{df(1)}$:	0	1.70	1.0	0	0	0	0	0	0	0
item 1	$S_{df(2)}$:	2.60	3.40	2.0	0	0	0	0	0	0	0
	$S_{df(3)}$:	1.30	0	1.0	0	0	0	0	0	0	0
	$S_{df(1)}$:	0	0	0	0	0	0	0	0.50	1.30	0
item 2	$S_{df(2)}$:	0	0	0	0	0	0	0	0	0	0
	$S_{df(3)}$:	0	0	0	0	0	0	0	0	0	0
	$S_{df(1)}$:	0	0	0	0	0	0	0.82	0	0	0
item 3	$S_{df(2)}$:	1.30	0	0	0	0	0	2.46	0	0	0
	$S_{df(3)}$:	0	0	0	0	0	0	0	0.50	0	0

If we consider $W_{df(1)} = 1.0$, $W_{df(2)} = 1.2$ and $W_{df(3)} = 0.8$, then the item vector S_i of these three items are given in Table 6.

Table 6. Calculated item vector S_i for above three items.

Final words	sport	athlet	Football	Row	Race	Wrestl	technolog	engin	applic	bus
S_1	3.64	4.42	3.8	0	0	0	0	0	0	0
S_2	0	0	0	0	0	0	0	0.5	1.3	0
S_3	1.04	0	0	0	0	0	2.788	0.6	0	0

Now, we can calculate the similarity between each item vector S_i and user vector U_v. To calculate the similarity, we can use any similarity measure such as the cosine similarity that makes more sense for our problem. If $a = (a_1, a_2, \dots a_n)$ and $b = (b_1, b_2, \dots b_n)$ are two n-dimensional vectors, then the cosine similarity between these two vectors, $\text{cosine}(\theta)$, is represented using a dot product and magnitude as:

$$Similarity(a, b) = cosine(\theta) = \frac{a.b}{\|a\|\|b\|} = \frac{\sum_{i=1}^{n} a_i b_i}{(\sum_{i=1}^{n} a_i^2)^{1/2} (\sum_{i=1}^{n} b_i^2)^{1/2}} \quad (2)$$

Based on above formula, the similarity between U_v and each S_i will be:

$W_1 = Similarity(U_v, S_1) = 0.61$; $W_2 = Similarity(U_v, S_2) = 0.23$; $W_3 = Similarity(U_v, S_3) = 0.73$

Finally, if the number of desired recommendations is $N = 2$, then we will recommend to the user the two items with the top W_i values, namely, item 3 and item 1.

2.3. Updating the user model (Learning process)

In this section, we introduce a method for updating the user model based on his/her history. Each user has some history data, including some negative and positive samples, where the negative samples are the items that the system recommended but the user did not like, and the positive samples are the items that the user liked. The process of updating the user model involves two major steps:

1) Updating the user model words.
2) Updating the user model weights.

2.3.1. Updating the user model words. In the previous section, we introduced how to create the initial user model for each user. In this section, we explain how our method updates the user model based on user history. Consider the sample initial user model as shown in Table 2.

For updating the user model words, we just need to add whatever words in all other users' models that do not exist in the current user model. We also consider weight=0 for any newly added word. For example, consider the above-mentioned user model and some other words from other users' models including "art", "dance", "artwork", "painting", "sculpture" and "artistry". Then, Table 7 shows the new user model.

Table 7. Updated model of the initial user model Table 2.

Word:	**Weight:**
sport	2.0
athletics	1.0
football	1.0
rowing	0.9
racing	0.9
wrestling	0.8
technology	2.0
engineering	0.8
IT	0.8
application	0.6
business	0.5
technological	0.5
art	2.0
dance	0
artwork	0
painting	0
sculpture	0
artistry	0

In the next section, we describe how we can update the weights of the created model.

2.3.2. Updating the user model weights. The process of updating the user model weights is:

1) Let the current user model be M_c, which is the output of the previous step (Updating the user model words).
2) For each rated item Q:
    a. Calculate the model of Q as M_Q (explained in the next paragraph)
    b. If Q was labeled as a negative sample, then $M_c = M_c - \alpha * M_Q$
    c. If Q was labeled as a positive sample, then $M_c = M_c + \alpha * M_Q$

Now, we explain how we can calculate the item model M_Q. This process is similar to the process of finding top items. However, there are some differences. Consider a positive or negative item Q:

a. For each data field j with significance weight $W_{df(j)}$:
    i. $S_{df(j)}$:= the tf-idf for the jth data field of item Q for the words that we have in the updated user model;
b. Calculate the item vector M_Q by weighted summation over all the n data fields:

$$M_Q = \sum_{j=1}^{n} [S_{df(j)} * W_{df(j)}] \quad (3)$$

3. Performance Evaluation

In this section, we evaluate experimentally the performance of our content-based recommendation system and various design decision choices. To do so, the college community is selected as a good sample of data-overwhelmed exclusive community to be used as a testbed. Therefore, our goal is to recommend some college events to students. All data that we used in these experiments is collected from the George Washington University website. However, we needed some of this data to be labeled by humans so we could compare the result of our method with labeled data.
To obtain this labeled data, we provided 10 different workbooks. Each workbook had 30 events. Then we gave workbooks to ten different users and asked them to provide:

1) Some keywords that indicate their hobbies, activities, skills, and issues that they might be interested in.
2) 10 events out of the 30 provided, that are more interesting to them.

3.1. Difference between stemming and lemmatizing

As we mentioned before, words might be in plural forms or even in other forms. To convert all forms of a word to the same word, we can use stemming or lemmatization. For a single input word, these two processes might result in different forms of the word. For example, consider word "technology"; the stemming process converts it to "technolog" and the lemmatizing process converts it to "technology".
In the first experiment, we are going to compare the following four alternatives: Original words, Stemmed words, Lemmatized words and, Union of Stemmed and lemmatized.

Note that the process of removing 'e', 'y' and 'i' letters from the end of the words is not performed in this experiment, but it will be carried out in the second experiment. Table 8 shows the results of matching for the four above situations. As you can see, using stemming produces higher precision and accuracy.

Table 8. Evaluation of our RS for when different versions of the words are used in calculations.

Method	Precision	Accuracy
Original words	0.66	0.77
Stemmed words	0.73	0.82
Lemmatized words	0.7	0.80
Union of Stemmed and Lemmatized words	0.7	0.80

3.2. Difference between keeping or removing 'e', 'y' and 'i' letters

In the course of this research, we found that removing the letters 'e' and 'y' from the end of lemmatized words could improve the efficiency of the method. The same improvement would be happened if we remove letter 'i' from the end of stemmed words. In the next experiment, we compare the performance of our method for when 'e', 'y' and 'i' are removed from the end of stemmed words vs. when they are not removed.

Table 9. Evaluation of our RS for when we use the process of removing 'e', 'y' and 'i' endings and for when we do not remove them

	Precision	Accuracy
Removing 'e' and 'y' (after lemmatizing)	0.72	0.81
Keeping 'e' and 'y' (after lemmatizing)	0.7	0.80
Removing 'i' (after stemming)	0.75	0.83
Keeping 'i' (after stemming)	0.73	0.82

As Table 9 shows, based on which version of the words (stemming or lemmatizing) is used in the previous step, removing 'e', 'y' and 'i' improves precision and accuracy of our method. Therefore, for the rest of the experiments, stemming will be used, and the letter 'i' will be removed from the end of the stemmed versions of words.

3.3. Difference between binary calculation or frequency calculation

As mentioned before, we used tf-idf for creating the user and item vectors. Actually, the tf-idf metric has two parts: term frequency (tf) and inverse document frequency (idf): *tf-idf* = $tf * idf$. In some situation, it's better to binarize tf, that is, set tf = 1 if $tf > 0$, and keep $tf = 0$ otherwise:

$$tf\text{-}idf = \begin{cases} idf & if \;\; tf > 0 \\ 0 & if \;\; tf = 0 \end{cases} \tag{4}$$

In this experiment, we want to evaluate our method for two alternatives:

- Frequency calculation: Calculating tf-idf based on its common definition.
- Binary calculation: Calculating tf-idf based on the binarized tf.

As you can see from the Table 10, the modified tf-idf method results in better precision and accuracy. We will use binary tf calculation in the remainder of the experiments.

Table 10. Evaluation of our RS for when the original definition of tf-idf is used and when the binarized tf is used.

	Precision	Accuracy
Frequency calculation	0.75	0.83
Binary calculation	0.78	0.85

3.4. Difference between different vector similarity measures

As we mentioned before, we should use a good similarity measure for computing the similarity between user vector and item vector. In the next experiment, we evaluate our method based on using different similarity measures. We compare four well-known similarity measures, namely, dot-product, cosine, Euclidean, and Manhattan. Table 11 shows the performance results.

Table 11. Evaluation of our RS for when we use different similarity measures.

Method	Precision	Accuracy	Method	Precision	Accuracy
Euclidean	0.62	0.75	Dot product	0.71	0.81
Manhattan	0.59	0.73	Cosine	0.78	0.85

As you can see from Table 11, the cosine similarity measure resulted in the highest precision and accuracy. Therefore, we will use this measure for calculating the similarity between two vectors.

3.5. Comparison of our method with some other methods

In this experiment, we compare our method with the following three methods of generating a feature-vector from text content.

1) TF/IDF,
2) *Word2vec* using Google pre-trained model [15]
3) *The glove method* using [Wikipedia 2014 + Gigaword 5] model [16].

Before presenting the results of this experiment, we explain how the last two above methods finds the top *N* recommended items for users:

- Word2vec Google trained model: Word2vec generates a vector for each word in an item text. The vectors of all the words in the text are added to create the feature of the item text.
- The glove method using [Wikipedia 2014 + Gigaword 5] model: It is similar to word2vec except that it uses different training algorithms.

All the above-mentioned methods including our method use the cosine measure for calculating the similarity between the user model and the item feature-vectors, and recommending the top *N* recommended items. In addition to the above methods, we compare the result of our method with UMBC phrase/sentence similarity service which is a well-known semantic textual similarity system. It gets two texts as input parameters and returns a real number as output which indicates the semantic similarity value between the two input texts. To calculate the semantic similarity between user interests and items, we feed the user keywords and the items text content as the first and second inputs of UMBC semantic similarity service respectively. Table 12 shows the result of comparison between our method and the above methods:

Table 12. Comparison between our method and some other methods.

Method	Precision	Accuracy	Method	Precision	Accuracy
Our method	0.78	0.85	Glove	0.68	0.79
TF/IDF	0.58	0.72	UMBC	0.73	0.82
Word2vec	0.67	0.78			

As you can see in table 12, although some models like the Google pre-trained model and the Glove pre-train model are strong, our method results in the best precision and accuracy. The reason for this is that these models are really general and cover so many words in the English language, whereas our problem is very specific and so our model only covers restricted vocabulary related to our problem, which leads to better results.

3.6. Comparison of our method before and after updating the user model

In this last experiment, we show how the updating process could improve the performance of our method. Table 13 shows the result of this experiment. As you can see, the precision and accuracy have been improved from 0.78 and 0.85 to 0.81 and 0.87 respectively.

Table 13. Evaluation of our method before and after training.

	Precision	Accuracy
Before training	0.78	0.85
After Training	0.81	0.87

4. Conclusion and future work

To overcome the data overload issue in recommendation applications, we proposed a novel text-content-based recommendation system and optimized it by considering different alternatives and different decision choices. The resulting system uses a new specific procedure to generate the user model and convert the text content of the community items to a meaningful feature-vector. The user model is generated by soliciting from a user a few keywords that represent his/her interests, and expanding those keywords into list of weighted near-synonyms. Since words might be various morphological and/or syntactical forms, we normalize all the forms of a word to the same word using stemming and/or lemmatization, and then remove 'e', 'y' and 'i' from the end of stemmed/lemmatized words. The item feature-vectors are generated from the textual descriptions of the items, using modified tf-idf values of the users' keywords and near-synonyms. Then by computing the cosine similarity between user and item feature-vectors, the items that best match the user's interests are recommended.

We tested the proposed method on a college community to recommend college events to students based on their interests. Experimental results show that stemming and removing 'i' letters from end of stemmed words increase the accuracy and precision significantly over any other combination (or lack) of stemming and lemmatization, that binarized-tf tf-idf is superior to non-binarized tf-idf, that the cosine similarity measure is superior to other similarity measures, and that model training (i.e., learning from past performance) improves the performance of the system. Also, we compared our system with some well-known methods like Glove and Word2Vec for creating the feature-vectors for text content, and found that our proposed procedure (of creating the user and item feature-vectors) resulted in highest precision (78%) and accuracy (85%) for our application among those competing methods.

As mentioned before, the hybrid recommendation systems could improve performance; however, to create the collaborative part of it, there need to be many pre-rated samples available. Currently there is no such data, but once there is, creating a collaborative or hybrid recommendation system will likely improve the performance of the proposed method. Also, creating and training a more knowledgeable model for computing more accurate near-synonyms will definitely improve the accuracy of recommendations.

References

[1] Gavalas D, Konstantopoulos C, Mastakas K and Pantziou G 2014 Mobile recommender systems in tourism *J. Netw. Comput. Appl.* **39** 319-33
[2] Bobadilla J, Ortega F, Hernando A and Gutiérrez A 2013 Recommender systems survey *Knowledge-based Sys.* **46** 109-32
[3] Ricci F, Rokach L and Shapira B 2011 *Introduction to recommender systems handbook* (US: springer) p 1
[4] Herlocker J L, Konstan J A, Terveen L G and Riedl J T 2004 Evaluating collaborative filtering recommender systems *ACM Trans. Inf. Syst.* **22** 5-53
[5] Lee W S 2001 June Collaborative learning for recommender systems *In ICML* **1** 314-21
[6] Adomavicius G and Tuzhilin A 2005 Toward the next generation of recommender systems: A survey of the state-of-the-art and possible extensions *IEEE Trans. Knowl. Data Eng.* **17** 734-49
[7] Aggarwal C C 2016 Content-based recommender systems *Recommender Sys.* 139-66
[8] Gu Y, Zhao B, Hardtke D and Sun Y 2016 April Learning global term weights for content-based recommender systems *25th Proc. Int. Conf. on WWW* 391-400
[9] Philip S, Shola P and Ovye A 2014 Application of content-based approach in research paper recommendation system for a digital library *Int. J. Adv. Comput. Sci. App.* **5** 37-40
[10] Lak P, Babaoglu C, Bener A B and Pralat P 2016 News article position recommendation based on the analysis of article's content-time matters *CBRecSys* 11
[11] Wintrode J, Sell G, Jansen A, Fox M, Garcia-Romero D and McCree A 2015 April Content-based recommender systems for spoken documents *IEEE Int. Conf. ICASSP* 5201-05
[12] Castro J, Rodriguez R M and Barranco M J, 2014 Weighting of features in content-based filtering with entropy and dependence measures *Int. J. Comput. Intell. Sys.* **7** 80-89
[13] Kamran Kowsari, Donald E. Brown, Mojtaba Heidarysafa, Kiana Jafari Meimandi, Laura E. Barnes, "Hdltex: Hierarchical deep learning for text classification," in Machine Learning and Applications (ICMLA), 2017 16th IEEE International Conference on. IEEE, 2017
[14] Kamran Kowsari, Maryam Yammahi, Nima Bari, Roman Vichr, Faisal Alsaby and Simon Y. Berkovich, "Construction of FuzzyFind Dictionary using Golay Coding Transformation for Searching Applications" International Journal of Advanced Computer Science and Applications(ijacsa), 6(3), 2015. http://dx.doi.org/10.14569/IJACSA.2015.060313
[15] Mikolov T, Chen K, Corrado G and Dean J 2013 Efficient estimation of word representations in vector space. *arXiv preprint arXiv* 1301.3781
[16] Pennington J, Socher R and Manning C D 2014 October Glove: global vectors for word representation. In EMNLP 14 1532-43

IOP Conf. Series: Materials Science and Engineering **261** (2017) 012018 doi:10.1088/1757-899X/261/1/012018

Comparisons and Selections of Features and Classifiers for Short Text Classification

Ye Wang[1, a], Zhi Zhou[2, b], Shan Jin[1,c], Debin Liu[2, d] and Mi Lu[1, e]

[1] Department of Electrical and Computer Engineering, Texas A&M University
[2]SocialCredits, Ltd

E-mail: [a]wangye0523@tamu.edu, [b]zhi.zhou@socialcredits.cn, [c]jinshan@tamu.edu, [d]debin.liu@socialcredits.cn, [e]mlu@ece.tamu.edu

Abstract. Short text is considerably different from traditional long text documents due to its shortness and conciseness, which somehow hinders the applications of conventional machine learning and data mining algorithms in short text classification. According to traditional artificial intelligence methods, we divide short text classification into three steps, namely preprocessing, feature selection and classifier comparison. In this paper, we have illustrated step-by-step how we approach our goals. Specifically, in feature selection, we compared the performance and robustness of the four methods of one-hot encoding, tf-idf weighting, word2vec and paragraph2vec, and in the classification part, we deliberately chose and compared Naive Bayes, Logistic Regression, Support Vector Machine, K-nearest Neighbor and Decision Tree as our classifiers. Then, we compared and analysed the classifiers horizontally with each other and vertically with feature selections. Regarding the datasets, we crawled more than 400,000 short text files from Shanghai and Shenzhen Stock Exchanges and manually labeled them into two classes, the big and the small. There are eight labels in the big class, and 59 labels in the small class.

1. Task and problem

Short text classification is unlike traditional long text documents, due to its own characteristics in terms of shortness and conciseness. The objective of this paper is to compare existing methods for better short-text classification.

Each short text is less than 20 words, and we manually labeled it into two class of tags, there are eight types of labels in the big class and fifty-nine types of labels in the small class. The total number of short text files is about 400,000, and each tag's short text files are equal. In other words, we can feed about 50,000 short text files to train each label in the big class, and around 6,700 files to train each label in the small class. There exist some challenges specific to short text classification. Shortness entails a lack of information and conciseness is a synonym for simplicity, which can cause confusion when we try to classify 59 labels with files of no more than 20 words. Besides, we have more than 400,000 short text files, which amount to approximately 5,349,348 words and would certainly create a lot of sparsity in the vectorization process. Sparsity would in turn lead to several problems such as data redundancy, sparsity matrix, etc. We will introduce them in next few sections.

Company Announcement	Big Class	Small Class
KPC Pharmaceuticals, Inc.: Announcement on the Progress of the Restricted Stock Grant in the 2016 Equity Incentive Plan	Major Events	Equity Incentive Compensation
ENN Ecological Holdings Co. Ltd.: 2016 Semi annual report	Financial Report	Semi Annual Report
BOETECHNOLOGYGROUPCO.,LTD: Announcement on repurchasing part of company shares and canceling the creditor	Equity and Capital Stock	Repurchasing Equity
Total: 409,871 short text announcement	8 labels	59 labels

Figure 1. Announcement examples in Dataset

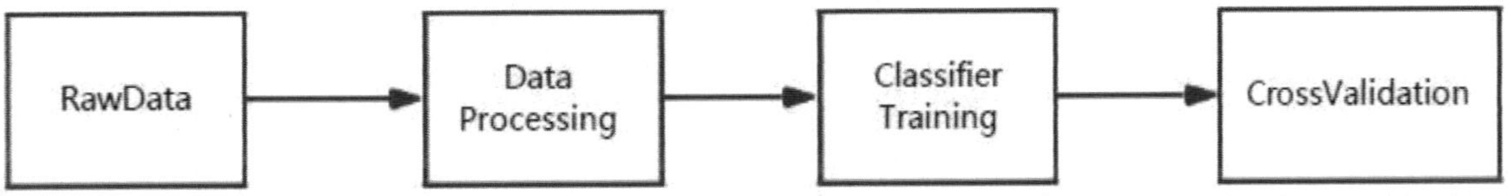

Figure 2. Basic Flow Chart

The rest of this paper is organized as follows: Section II presents a high-level description of the approach in terms of data processing and classifier training. Section III introduces the system implementing our task and how we handle the challenges. Section IV discusses the experiment results, and the paper concludes with Section V.

2. High-level description of approach

As can be seen, Figure 1 lists some examples of our data samples which will be used later. Figure 2 shows the basic flow chart, which can be regarded as top level description of our approach. In this section, we will discuss in details about the entire process.

2.1. Data processing

There are many different data processing techniques. We should be aware of their characteristics and choose the appropriate method.

2.1.1. SEGMENTATION. According to [1], segmentation is becoming increasingly more important in Chinese, Japanese and many other Asian language processing tasks. Unlike English, Chinese words are not delimited by whitespace characters, so word segmentation is a fundamental first step in processing these languages. Several algorithms have been proposed for Chinese word segmentation [2], and the study in automatic Chinese word segmentation has made significant progress in recent years. For the purpose of our study, we just chose the currently most popular segmentation method which is based on prefix trie and the Viterbi algorithm.

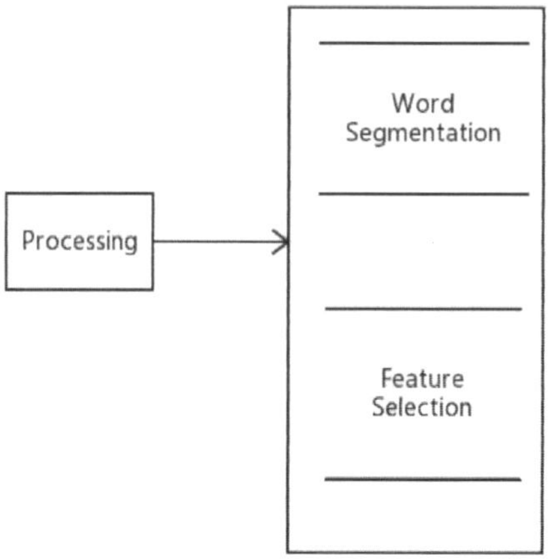

Figure 3. Processing the Data

2.1.2. FEATURE SELECTION. Figure 3 shows the data processing flow chart. Once we get the segmented words, we can convert them into a vector matrix for later training. This process is called word embedding, or distributional models. The reason for constructing such a vector matrix is that we can utilize the term-context matrix to represent the short text, which is much simpler for training purposes. There are many ways to construct the vectors, such as sparse vectors and dense vectors. Sparse vectors have most elements equal to zero and lengths of about 20,000 to 50,000, which will be very time-consuming computationally, while dense vectors are constructed in 100-500 dimensions, so are much faster than sparse vectors when used in training and classifications. Dense vectors may also better capture synonymies than sparse vectors use [3]. Moreover, we employed two methods for the sparse vector construction, i.e. counter vectorizer and term frequency-inverse document frequency (tf-idf). Counter vectorizer is also called one-hot coding, which is applied to categorical features.

Categorical features are "attribute-value" pairs where the value is restricted to a list of discrete possibilities without ordering. Counter vectorizer is like a raw vectorizer, while tf-idf is more refined, since it is a numerical statistic that is intended to reflect how important a word is to the short text in our collections. It is often used as a weighted factor in application. The key characteristic of tf-idf is that it increases proportionally with the frequency of a word appearing in the document, but is offset by the frequency of the word in the corpus, which helps to adjust for the fact that some words appear more frequently in general [4].

Compared with sparse vectors, dense vectors are more popular because they are shorter and yet more meaningful. There are three popular methods for constructing dense vectors, i.e. singular value decomposition, neural language models and Brown clustering. Here we would like to focus on the neural language model, which is the state-of-the-art method for dense vector construction [3] [5]. There are two type of neural language models, Skip-gram and continuous bag of words (CBOW), which are also collectively called word2vec models, as shown in Figure 4. [6] provides a detailed explanation of the two models. In short, the CBOW architecture predicts the current word based on the context while skip-gram predicts surrounding words given the current word. One advantage of dense vectors is that we can get a short and yet meaningful vector to represent each word. Also, it has the superior characteristic that the matrix of similar words also has a closer distance, which is helpful to constructing the thesaurus. Moreover, [7] enhanced his work of word embedding and proposed a novel model called paragraph2vec (or doc2vec) as shown in Figure 5. This method trains the entire document as a vector matrix, while for word2vec, the basic idea is to predict the word. Similar to word2vec, in the training of the document vector we need to go through the whole text. We have also implemented doc2vec in this paper and will compare it with other feature selection methods in the results and analyses part.

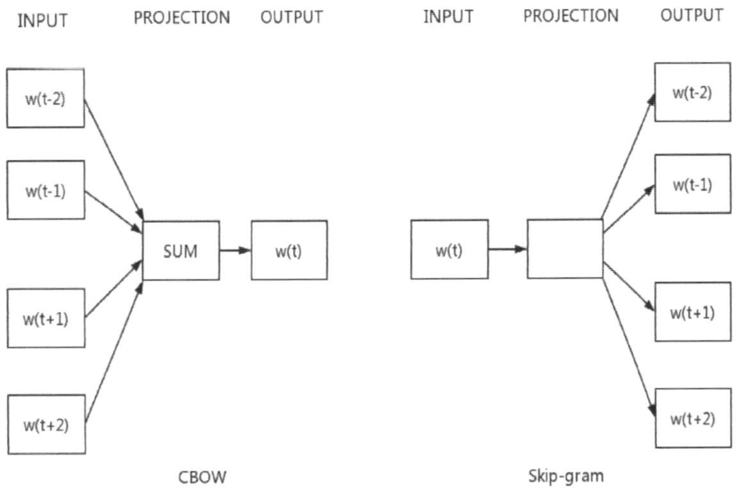

Figure 4. Two Models of word2vec

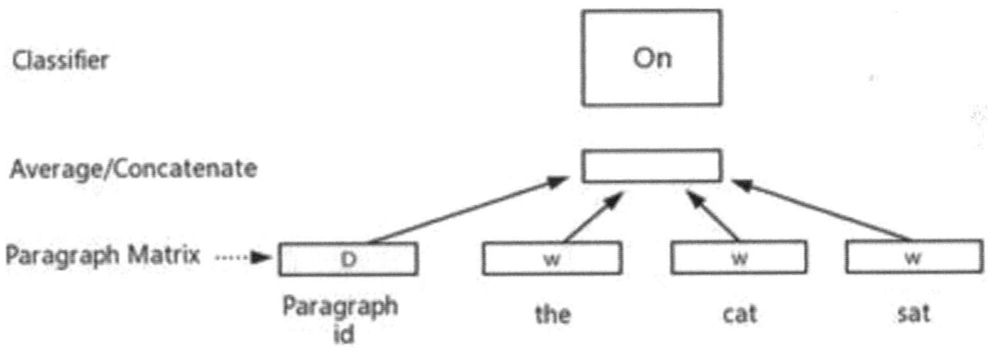

Figure 5. Paragraph2vec Model

2.2. CLASSIFICATION

Once the feature is selected, its time to train the classifier. Classification is one of the most important steps in all machine learning's tasks. Classification is the problem of identifying to which set or category a new observation belongs, on the basis of a training set of data containing observations whose class is known. Since we already have labeled all the instances, we only need to choose supervised learning classifiers. Among the various learning algorithms, we cannot simply decide the best one before experimenting and comparing some of them. Hence we selected several popular classifiers, including naive Bayes (NB), decision tree (DT), k-nearest neighbor (KNN), logistic regression (LR) and support vector classifier (SVC), and applied them with both the big and small classes of labels. Each classifier has its own advantages and disadvantages.

2.2.1. NAIVE BAYES. Naive Bayes methods build upon the famous Bayes' theorem with the (not so) "naive" assumption of independence between each pair of features. The NB method is very low time complexity, and its assumption usually works quite well in some real-world situations such as spam filtering and document classification. As a consequence of the decoupling of the conditional probability distributions of different features, the probability distribution of each feature can be independently estimated as one-dimensional distribution, which in turn helps alleviate problems stemming from the curse of dimensionality. In this study we compared both Gaussian (GNB) and multinomial (MNB) naive Bayes classifiers. When dealing with dense vectors, we treat those data as continuous data, and when dealing with continuous data, a typical assumption is that the continuous values associated with each class are distributed according to a Gaussian distribution.

2.2.2. LOGISTIC REGRESSION. Multinomial logistic regression is known by a variety of other names, including polytomous LR, multi-class LR, softmax regression, multinomial logit, maximum entropy (MaxEnt) classifier, and conditional maximum entropy model. In fact, multinomial logistic regression is a classification method that generalizes logistic regression to multiple-class problems. In this model, the probabilities describing the possible outcomes of a single trial are modeled using a logistic function.

2.2.3. SUPPORT VECTOR MACHINE. Support Vector Machine (SVM) is widely used among classification, regression and even outlier detection. The advantage of SVM is obvious: First, it is very effective not only in high dimensional spaces, but also in cases whether the number of dimensions is greater than the number of samples. Second, it is considerably memory efficient due to its own advantage of kernel mapping to high-dimensional feature spaces [8]. Just as every coin has two sides, the disadvantage of SVM is that if the number of feature is much greater than the number of samples, the method is likely to give poor performance. A linear support vector classifier (SVC) is used in this paper.

3. SPECIFIC SYSTEM IMPLEMENTATIONS

The whole structure of our system is divided into four parts as we illustrated before: Getting raw data, processing data, training classifier and cross validation. We implemented it by Python2.7 with some open source APIs like Scikit-learn [9] and Gensim [10].

4. EXPERIMENT

Several classifiers have been trained to classify Chinese short text les, including GNB, MNB, SVC, LR, KNN and DT [11]. However, we will only present the experiment results of MNB, SVC and LR in this section, since they are much better than those of the other classifiers.

Figure 6 to Figure 10 show the 5-fold cross-validation results for each of the three classifiers. In the cross-validation tests we have used four features, i.e. word2vec, doc2vec, tf-idf and counter vectorizers. The performances of the last two are similar so we treat them indifferently as a single tf-idf/counter feature. In addition, we have filtered stop words in each experiment, which also slightly improves the accuracy.

From the tables we can see that in all cases, the tf-idf/counter feature has the highest accuracy, while word2vec next, and doc2vec the lowest. The feature doc2vec produces the worst result in any circumstance, even much worse than plain guessing, which is different from the long text classification results [5]. We also get different results for the big and small classes of labels. The small class generally results in higher accuracy than the big class, which is counterintuitive and needs further investigation.

IOP Conf. Series: Materials Science and Engineering **261** (2017) 012018 doi:10.1088/1757-899X/261/1/012018

big				small			
Data = ½				Data = ½			
Classifier	Accuracy, %			Classifier	Accuracy, %		
	word2vec	doc2vec	tf-idf/counter		word2vec	doc2vec	tf-idf/counter
MNB	46.5454798	30.979916	67.9974846	MNB	50.5129724	3.9339202	71.8358194
SVC	56.523723	20.7917638	70.6630796	SVC	77.1284876	2.055327	83.5525026
LR	64.3266498	30.979916	70.7194048	LR	81.4567048	2.4392982	84.218733
Data = 1/5				Data = 1/5			
Classifier	Accuracy, %			Classifier	Accuracy, %		
	word2vec	doc2vec	tf-idf/counter		word2vec	doc2vec	tf-idf/counter
MNB	46.020836	30.9789578	66.8190094	MNB	47.8368036	3.6108782	71.3708568
SVC	58.7445636	22.0106284	69.811158	SVC	70.3710892	2.7923356	83.9617922
LR	64.9365838	30.9789578	70.0025738	LR	81.5665502	2.438548	84.1286016
Data = 1/10				Data = 1/10			
Classifier	Accuracy, %			Classifier	Accuracy, %		
	word2vec	doc2vec	tf-idf/counter		word2vec	doc2vec	tf-idf/counter
MNB	45.997871	30.979801	64.6257898	MNB	47.0733842	3.5133214	71.528666
SVC	56.9952578	25.7708636	69.0387578	SVC	60.1547316	2.1768262	83.9285676
LR	64.455278	30.979801	68.6656358	LR	81.3346074	2.4397412	83.5524258

Figure 6. Result

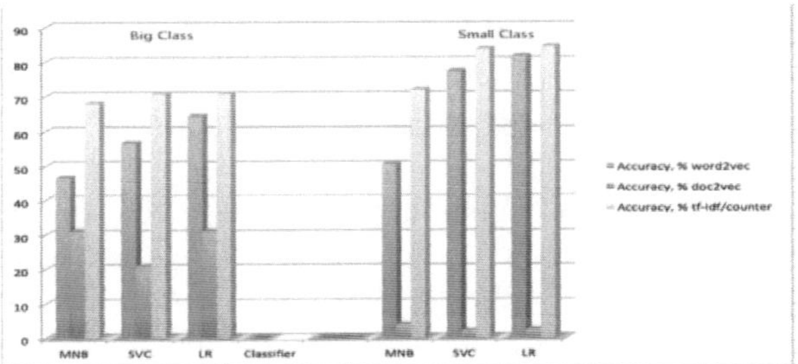

Figure 7. 1/2 datasets of overall

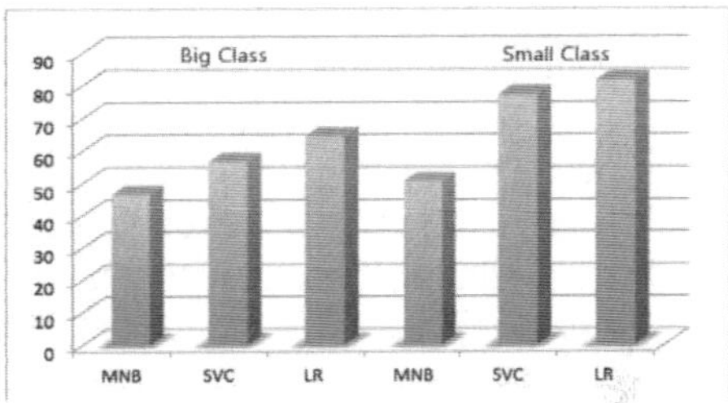

Figure 8. 1/2 datasets of word2vec

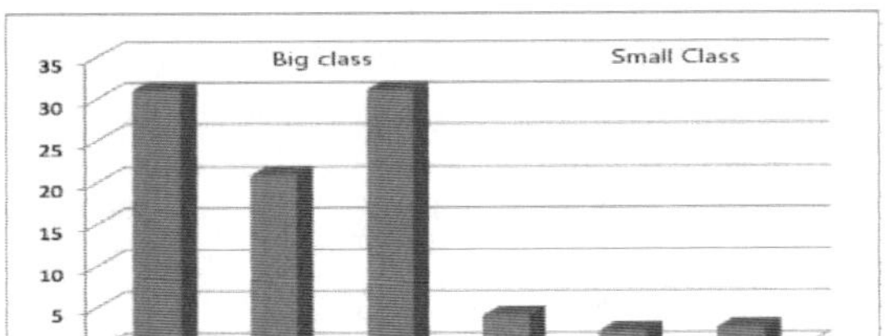

Figure 9. 1/2 datasets of doc2vec

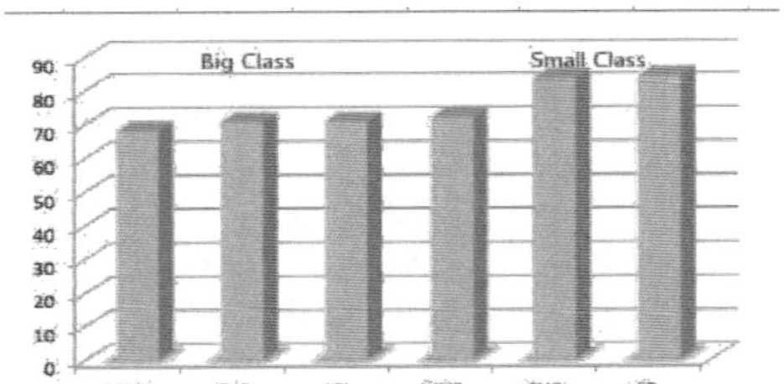

Figure 10. 1/2 datasets of tf-idf/counter

A comparison of different classifiers would show that with the tf-idf/counters feature, LR and SVC are much better than MNB, and the results of the two are comparable with each other. This may not be the case when other features are used. While a high accuracy is expected for the SVC because of the use of kernel function, it is a little surprising that the overall highest accuracy 84.22% is associated with LR.

Additionally we have tried to change the size of the dataset, and it seems increasing the size of the dataset and raises the accuracy, but this impact is not significant.

5. CONCLUSIONS

In this paper, we have demonstrated the classification of Chinese short text, in the context of public financial documents. Different features and classifiers are applied and compared, and the cross-validation results show that logistic regression and support vector classifier with the tf-idf or CounterVectorizer feature attain the highest accuracy and are the most stable in all circumstances. We have also observed some distinct characteristics of the short text classification problem like that the small class produces better results than big class, and that doc2vec always doesn't work well.

References

[1] Blei, D.M., Ng, A.Y., Jordan, M.I.: Latent dirichlet allocation. Journal of machine Learning research 3(Jan), 993{1022 (2003)

[2] Buitinck, L., Louppe, G., Blondel, M., Pedregosa, F., Mueller, A., Grisel, O., Niculae, V., Prettenhofer, P., Gramfort, A., Grobler, J., Layton, R., VanderPlas, J., Joly, A., Holt, B., Varoquaux, G.: API design for machine learning software: experiences from the scikit-learn project. In: ECML PKDD Workshop: Languages for Data Mining and Machine Learning. pp. 108-122 (2013)

[3] Hand, D.J., Mannila, H., Smyth, P.: Principles of data mining. MIT press (2001)

[4] Huang, C., Zhao, H.: Chinese word segmentation: A decade review. Journal of Chinese Information Processing 21(3), 8{20 (2007)

[5] Le, Q.V., Mikolov, T.: Distributed representations of sentences and documents. In: ICML. vol. 14, pp. 1188{1196 (2014)

[6] Maji, S., Berg, A.C., Malik, J.: Classification using intersection kernel support vector machines is efficient. In: Computer Vision and Pattern Recognition, 2008. CVPR 2008. IEEE Conference on. pp. 1{8. IEEE (2008)

[7] Mikolov, T., Chen, K., Corrado, G., Dean, J.: Efficient estimation of word representations in vector space. arXiv preprint arXiv:1301.3781 (2013)

[8] Mikolov, T., Sutskever, I., Chen, K., Corrado, G.S., Dean, J.: Distributed representations of words and phrases and their compositionality. In: Advances in neural information processing systems. pp. 3111-3119 (2013)

[9] Peng, F., Feng, F., McCallum, A.: Chinese segmentation and new word detection using conditional random elds. In: Proceedings of the 20th international conference on Computational Linguistics. p. 562. Association for Computational Linguistics (2004)

[10] Rehurek, R., Sojka, P.: Software Framework for Topic Modelling with Large Corpora. In: Proceedings of the LREC 2010 Workshop on New Challenges for NLP Frameworks. pp. 45{50. ELRA, Valletta, Malta (May 2010), http://is.muni.cz/publication/884893/en

[11] Rokach, L., Maimon, O.: Data mining with decision trees: theory and applications. World scientific (2014)

IOP Conf. Series: Materials Science and Engineering **261** (2017) 012019 doi:10.1088/1757-899X/261/1/012019

Research on Long Tail Recommendation Algorithm

Xuezhi Hu[1,a], Chuang Zhang[1,b], Ming Wu[1,c] and Yang Zeng[1]

School of Information and Communication Engineering, Beijing University of Posts and Telecommunications
Xitucheng Road No.10, Beijing, P.R.China

[a]huxuezhi@bupt.edu.cn; [b]zhangchuang@bupt.edu.cn; [c]wuming@bupt.edu.cn

Abstract. Most recommendation systems in the major electronic commerce platforms are influenced by the long tail effect more or less. There are sufficient researches of how to assess recommendation effect while no criteria to evaluate long tail recommendation rate. In this study, we first discussed the existing problems of recommending long tail products through specific experiments. Then we proposed a long tail evaluation criteria and compared the performance in long tail recommendation between different models.

1. Introduction

Recommender systems enable recommendations to be made to users of a system in reference to the items or elements. Currently, collaborative filtering [1] is the most commonly used and studied technology. One difficult, though common, problem for a recommender system is that most of the recommended information is concentrated in a small number of population items, which we called “Long tail phenomenon” [2][3]. Hence we proposed a detailed long tail evaluation criteria including long tail recommendation rate, long tail stability and depth which satisfies the quantitative evaluation of long tail recommendation.

In this paper, we first carried out a detailed feature engineering work. Then we compared GDBT model [4] and LR model to evaluate the performances in recommendation. The result shows that GBDT model achieves a better performance. In order to measure the utilization rate and overall performance of the long tail data, we proposed a new evaluation criteria system and also compared two traditional models with a long tail recommendation algorithm based on user-entropy LDA model [5][6][7], which improves the long tail recommendation rate and stability ensuring the traditional recommended performance not significantly reduced.

2. Data feature engineering

2.1. Feature extraction

The dataset we use in this paper comes from the user behavior log of Tmall, an E-commerce platform in 200 days, which contains the basic information of users and items, also the user behaviors. The amount of the data is over 200,000 and the total size of data is 700MB. In the initial research data, the information mainly includes users(user_id), purchase goods, brands(brand_id), category information (cat_id), behavior date(date), and types of user behaviors(click_cat) and so on.

According to the characteristics of Tmall data, we first divide the features into three categories. The first category is information of user characteristics, including the number of active days, the frequency of purchases, the total number of purchases and the conversion rate, which mainly reflects the influence of the user entity itself on E-commerce platform. The second category is information of behavior characteristics, including the average number of visits, the number of corresponding behaviors, and the days of these behaviors, which reflects the behaviors of users in using E-commerce platform. The last category is bag-of-words information, including browsing brands, categories of sellers and time, which mainly reflects the information of the platform itself.

After the preliminary analysis of the data features, as is shown in Table 1, we found that the amount of purchase goods is too much to cluster which is unable to extract the valid feature information. In the meantime, the user behaviors such as "Add to cart" can be ignored.

Table 1. F1 value of different models.

Feature types	Amount	User behaviors	Amount
Purchase goods	61553	Click	10928941
Types of goods	1226	Add to cart	41
Sellers	4275	Purchase	751502
Brands	6304	Favorite	682479

According to the data model given by the research object, we carried on detailed extraction including population preference and date of behaviors analysis. With the statistical characteristics of timing sequence, we can extract the characteristics of special festivals and cycle of crowd behaviors, inferring the basic cycle and time nodes of different people shopping.

Research on crowd preference is also an important part of feature extraction. We extracted some special brands and analyzed after normalization. The result shows a large number of brands do have the characteristics of the crowd preference. As is shown in Figure 1, the distribution of purchase or click are more concentrated on specific crowds. After screening the feature information, a total of 770 dimension features were extracted, including brand characteristics, seller behaviors, user behaviors, etc.

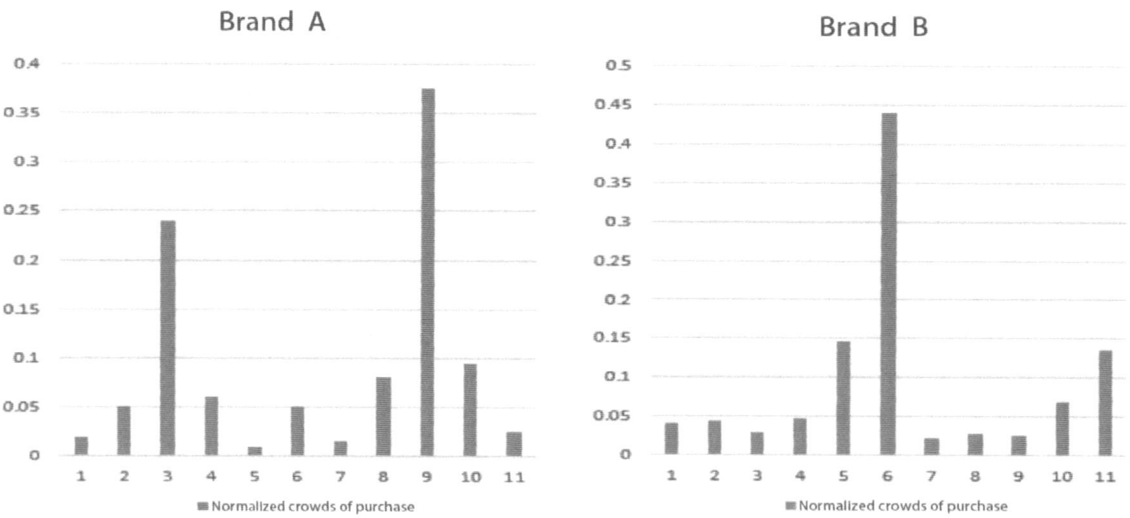

Figure 1. Statistics of purchase behavior in different crowds

2.2. *Feature selection*

The features are dimensioned by the calculation of similarity measure to achieve the purpose of dimensionality reduction, considering the correlation and redundancy between the features. We chose the Pearson correlation coefficient [8] as the relevance metrics. The specific approach is to use the Pearson correlation coefficient to estimate the correlation of each feature dimension. We reduce the dimension of the features by setting appropriate threshold, and finally select the relatively high correlation characteristics as the main features to train the model.

The Pearson correlation coefficient is defined as the ratio between the covariance and the standard deviation of random variables X and Y:

$$\rho_{X,Y} = \frac{Cov(X,Y)}{\sigma_X \sigma_Y} = \frac{\sum_{i=1}^{n}(X_i - \bar{X})(Y_i - \bar{Y})}{\sqrt{\sum_{i=1}^{n}(X_i - \bar{X})^2}\sqrt{\sum_{i=1}^{n}(Y_i - \bar{Y})^2}} \quad (1)$$

We defined the user behaviors as variable X and the multidimensional features as variable Y. The user behaviors include click, favorite, add to cart and purchase. By calculating the correlation user behaviors and the features, the Pearson correlation coefficient is gained. Part of the experimental results are shown in Table 2. The features of the final input data set were reduced to 50 dimensions using Pearson correlation coefficient to evaluate the relativity of each dimension. They are divided into three categories to characterize the entire model, as is shown in Figure 2.

Table 2. Part of Pearson correlation coefficient results

	Active days	**Average pre-purchase clicks**	**Conversion rate**	**Browse categories**	**Purchase rates**	**Retention rates**
Click	-0.148	0.074	0.011	-0.103	0.112	0.078
Add to cart	0.012	-0.108	0.092	-0.067	0.103	-0.131
Purchase	0.078	0.186	-0.156	0.078	-0.082	0.054
Favorite	-0.053	0.153	0.12	-0.065	-0.197	-0.03

3. Long tail evaluation criteria

3.1. Related works

Considering the high solution speed of LR model and the comprehensiveness of GBDT model, we discussed two fusion programs. A LDA model based on user-entropy is also mentioned.

3.1.1. Cascade model. We use GBDT to build two types of trees, non-behavior features tree and behavior features tree [9]. The non-behavior features tree is base which means non-behavior features can be based on this model to get the corresponding information. The behavior features tree is used to identify the features and feature combinations corresponding to behaviors. The mapping relations are shown in Figure 3. When a sample X comes in, the resulting feature is input to LR after traversing two types of tree-to-leaf nodes, and the non-behavior feature tree can be used as a supplemental tree.

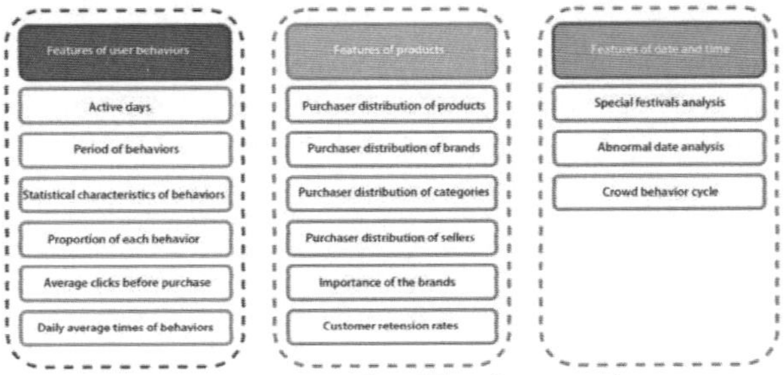

Figure 2. Feature classification results

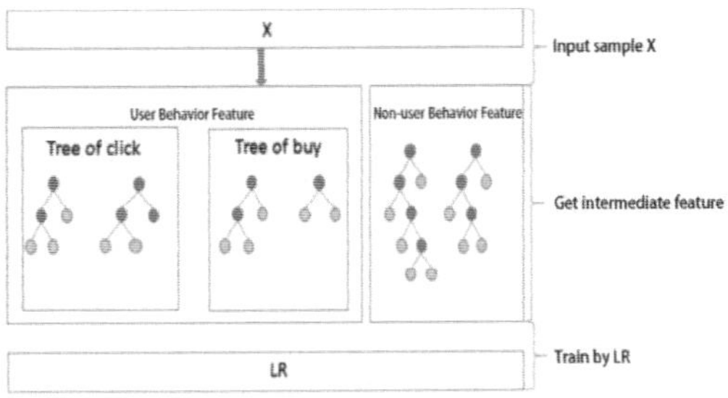

Figure 3. Cascade model processing

3.1.2. Artificial fusion. First of all, we extract N users from user set A to train the GBDT model, then split out M users to recommend using this model. We retrain the inaccurate part using LR model, the inaccurate part in the second part will be trained using GBDT model again. Repeat the above steps until all the users in the data set are removed.

The result is shown in Table 3. We use F1 value [10] to represent the performance of recommendation. It shows the cascade model performs better than any other models. However, according to statistics, only 30% of the recommended items come from the long tail part (We classified the items in proportion to 8:2 and the small part is long tail part), which means the existing evaluation indexes cannot evaluate the long tail characteristics.

Table 3. F1 value of different models.

	LR	GBDT	Artificial Fusion	Cascade Model
F1	0.312	0.321	0.319	0.336

3.1.3. EN-LDA model. The LDA model is a method of modeling the subject information in the text data [11]. It assumes that the document set can be divided into a number of implicit topics, and the topology of these topics is linear. Using the probabilistic inference algorithm, a single document can be represented as a mixture of these specific topics.

We use the product scores to measure the strength of the relationship between the user and the goods. A user's preferences for different items may not be the same, for example, he may give a high score to his favorite goods. We use $w(u,i)$ to represent the score of $\text{item}_{u,i}$ ranks *i*. In the algorithm, as is shown in Figure 4 [2], $w(u,i)$ represents the frequency of appearance in goods set S_u, θ represents the theme distribution of each user, and ψ represents the distribution of each good.

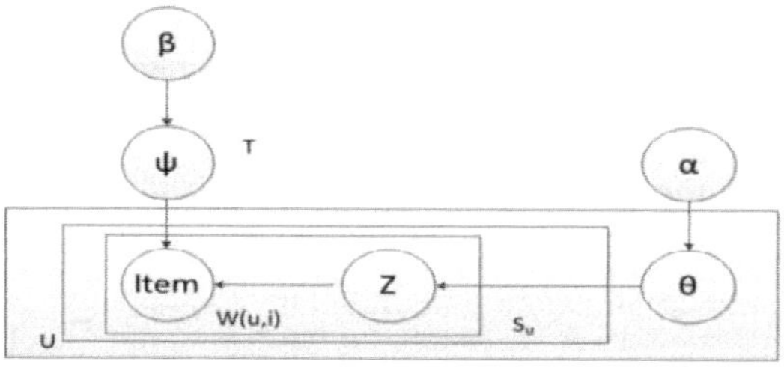

Figure 4. LDA model

The parameters of the EN-LDA [2] model are determined by using Gibbs sampling [12], and finally give the distribution of each user's theme. Use Gibbs sampling to modify the subject distribution iteratively until the model parameters converge.

3.2. Long tail recommendation rate

Long tail recommendation rate refers to the percentage of recommended products in the "long tail" part after classifying according to 8:2 way [13] in unit time.

$$\text{Longtail} = \frac{item_l}{item_{all}} \tag{2}$$

where $item_l$ represents the number of products in long tail, and $item_{all}$ represents the total number of recommendation. Long tail recommendation rate describes the long tail data utilization in recommendation systems.

3.3. Long tail stability

In addition to considering the long tail performance, the main task of the recommended system is to recommend the most reliable and practical product for the user. Based on this, we present the recommendation stability:

$$\text{Stability} = \frac{Longtail}{Accuracy} + Accuracy \tag{3}$$

where *Longtail* represents the long tail recommendation rate and *Accuracy* represents the accuracy of recommendation. The long tail recommended stability represents the balance between recommendation performance and long tail utilization, which comprehensively reflects the recommended metrics for the system.

3.4. Long tail depth and long tail level

Long tail depth is mainly used to describe the state of current system using the long tail data. As shown in Figure 5, we divide the long tail depth into three stages: initial long tail, intermediate long tail and deep long tail. The long tail level is divided into three according to the value of long tail recommended rate. Table 4 shows the definition of long tail.

Table 4. Definition of long tail

Conditions	Long tail level
Recommendation rate of deep long tail ≥ 10%	Third level
Not third level & Recommendation rate of intermediate long tail ≥ 10%	Second level
Not second level & Recommendation rate of initial long tail ≥ 10%	First level
Others	Non-Longtail

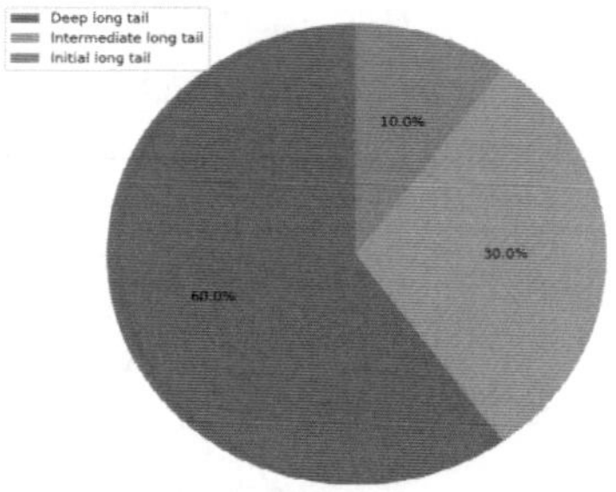

Figure 5. Long tail depth distribution

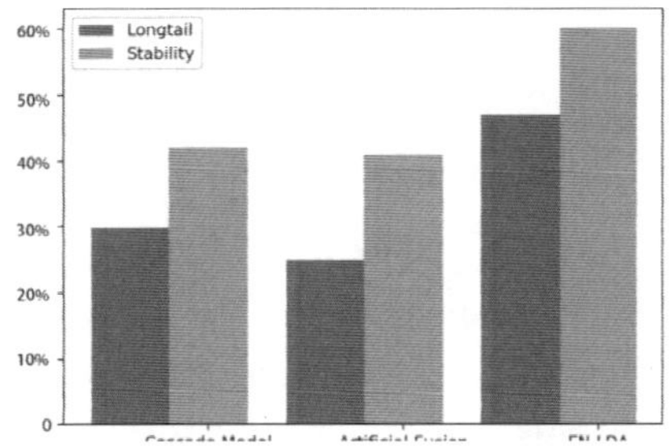

Figure 6. Long tail indicator between different models

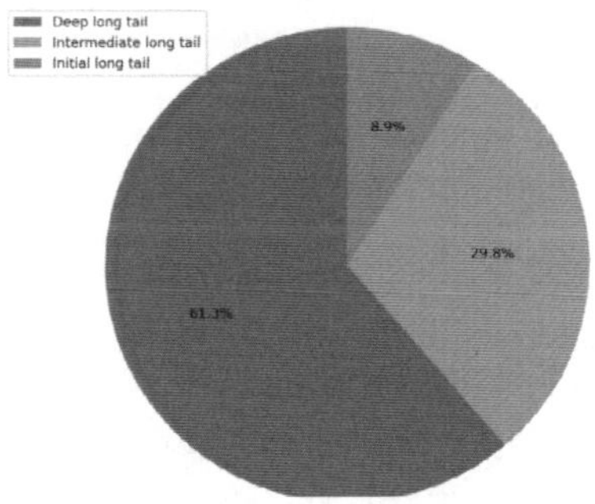

Figure 7. Long tail depth of recommended items

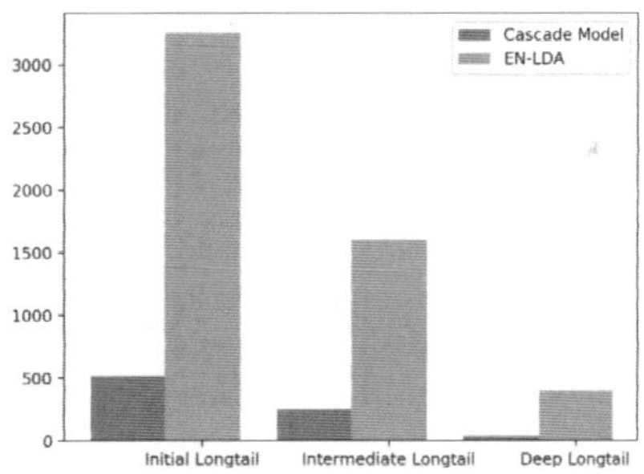

Figure 8. Long tail depth between different models

4. Experimental Results

In the previous chapter, we present two evaluation indicators of the long tail attributes: long tail recommendation rate (Longtail) and stability (Stability). The experiments in this chapter hope to extend the recommended range to the long tail by the subject model.

Figure 6 shows the comparison of the long tail indicator between the cascade model and the EN-LDA model. In the case of the same data set and other indicators are optimal, the long tail recommendation rate of the EN-LDA model is about 47.6% and the stability is about 60.1% which are both higher than other models. Ensuring that the recommended effect is not greatly affected, EN-LDA model can dig out the value of long tail products as much as possible, proving the feasibility of the subject model for long tail recommendation.

At the same time, we carefully divide the long tail stages of the recommended products, as is shown in Figure 7. It can be seen from the figure that long tail recommended goods are mainly concentrated in the initial long tail stage, accounting for more than 60%; intermediate long tail and deep long tail distribution is about 30% and 10%. While comparing to cascade model in Figure 8, the overall utilization of the cascade model is much smaller than the EN-LDA model, indicating that the EN-LDA model is more comprehensive in considering long tail data.

5. Conclusion

Our research is dedicated to solving the long tail problems of traditional recommendation algorithms. In order to ensure the research is based on real scene, we chose the user data of Tmall platform. We first divided the main features into behaviour features, product features and time features through extraction and selection. Then, we showed the long tail problem through experimental comparison of cascade models. In order to further quantify the problem, we proposed a detailed long tail evaluation criteria. And finally we validated the long tail criteria in evaluating the recommended diversity by comparing the performance of the EN-LDA model and the cascade model on this criteria.

On the basis of the traditional classification regression algorithm, the EN-LDA model uses the idea of the theme model to refine the mapping of the LDA algorithm, mapping the breadth of the theme to the long tail data diversity, and tries to solve the long tail problems. From the results of the experiment, it can be seen that the LDA model shows a very strong performance in the long tail evaluation criterion. The indicators have verified the model for long tail data hypersensitivity and high utilization, reflecting that recommendation and long tail phenomenon are not two irreconcilable opposition application scenarios.

References

[1] Silva E Q D, Camilo-Junior C G, Pascoal L M L, et al. An evolutionary approach for combining results of recommender systems techniques based on collaborative filtering[J]. Expert Systems with Applications, 2016, 53:204-218.
[2] Yin H, Cui B, Li J, et al. Challenging the Long Tail Recommendation[J]. Proceedings of the Vldb Endowment, 2012, 5(9):896-907.
[3] Brynjolfsson E, Hu Y J, Smith M D. From Niches to Riches: Anatomy of the Long Tail[J]. Social Science Electronic Publishing, 2006, 47(4):67-71.
[4] Yao L, Sheng Q Z, Segev A, et al. Recommending Web Services via Combining Collaborative Filtering with Content-Based Features[C]// IEEE, International Conference on Web Services. IEEE, 2013:42-49.
[5] Wei X, Croft W B. LDA-based document models for ad-hoc retrieval[J]. 2006:178-185.
[6] She J, Chen L. TOMOHA: Topic model-based Hashtag recommendation on twitter[C]// International Conference on World Wide Web. ACM, 2014:371-372.
[7] Richardson M, Dominowska E, Ragno R. Predicting clicks:estimating the click-through rate for new ads[C]// International Conference on World Wide Web. ACM, 2007:521-530.
[8] Benesty J, Chen J, Huang Y, et al. Pearson correlation coefficient[M]//Noise reduction in speech processing. Springer Berlin Heidelberg, 2009: 1-4.
[9] Tang X, Yang C C. TUT: a statistical model for detecting trends, topics and user interests in social media[M]. ACM, 2012.
[10] Ma H, Yang H, Lyu M R, et al. SoRec:social recommendation using probabilistic matrix factorization[C]// ACM Conference on Information and Knowledge Management, CIKM 2008, Napa Valley, California, Usa, October. DBLP, 2008:931-940.
[11] Sun C, Gao B, Cao Z, et al. HTM: a topic model for hypertexts[C]// Conference on Empirical Methods in Natural Language Processing. Association for Computational Linguistics, 2008:514-522.
[12] Lambiotte R, Salnikov V. Effect of memory on the dynamics of random walks on networks[J]. Eprint Arxiv, 2014, 3(2):1373–1378.
[13] Brynjolfsson E, Simester D. Goodbye Pareto Principle, Hello Long Tail: The Effect of Search Costs on the Concentration of Product Sales[J]. Social Science Electronic Publishing, 2011, 57(8):1373-1386.

IOP Conf. Series: Materials Science and Engineering **261** (2017) 012020 doi:10.1088/1757-899X/261/1/012020

Recognition of Bullet Holes Based on Video Image Analysis

Zhu Ruolin [1, 2], **Liu Jianbo** [1], **Zhang Yuan** [1] **and Wu Xiaoyu** [1]

[1]School of Information Engineering, Communication University of China, Beijing 10024, China

[2]zhuruolin@cuc.edu.cn

Abstract. The technology of computer vision is used in the training of military shooting. In order to overcome the limitation of the bullet holes recognition using Video Image Analysis that exists over-detection or leak-detection, this paper adopts the support vector machine algorithm and convolutional neural network to extract and recognize Bullet Holes in the digital video and compares their performance. It extracts HOG characteristics of bullet holes and train SVM classifier quickly, though the target is under outdoor environment. Experiments show that support vector machine algorithm used in this paper realize a fast and efficient extraction and recognition of bullet holes, improving the efficiency of shooting training.

1. Introduction

Machine learning is a research focus in the field of AI(artificial intelligence) currently. Video Image Analysis based on machine learning is applied in a variety of fields recently, such as biomedical image analysis, vehicle target recognition, expression recognition and character recognition. This paper is aimed at extracting and recognizing Bullet Holes from videos about the training of military shooting. Video Image Analysis can get the distribution of the point of impact during the shooting training of soldiers, and can remind us to take suitable training means to improve the military shooting level of soldiers.

Recognition of bullet holes in video has not only the advantages of high precision, low cost etc, but also the challenging owing to the background region and the noisy stream[1]. Nowadays the recognition of bullet holes mostly uses shallow machine learning methods, and normally their research objects is Chest round target. Liu Q and Chen Y[2] have calculated the position of bullet holes based on threshold segmentation and Boosted Cascade algorithm. Recently, deep learning become hot topics in Video Image Analysis, because of its strong ability to extract robust features. Deep learning has great potential in the field of image recognition.

This paper studies the image recognition of bullet holes based on shallow machine learning and deep learning method. According to the actual shooting situation, we design two algorithms to recognize bullet holes. The first one is based on rough positioning, Histogram of Oriented Gradients (HOG) and Support vector machine(SVM)[3]. The other applies rough positioning and deep convolution neural network method for comparison experiments. Both algorithms require rough positioning of Bullet holes, which contains extracting frames, Otus algorithms, morphological filtering and region labeling process. The process of bullet holes positioning is shown in Figure 1.

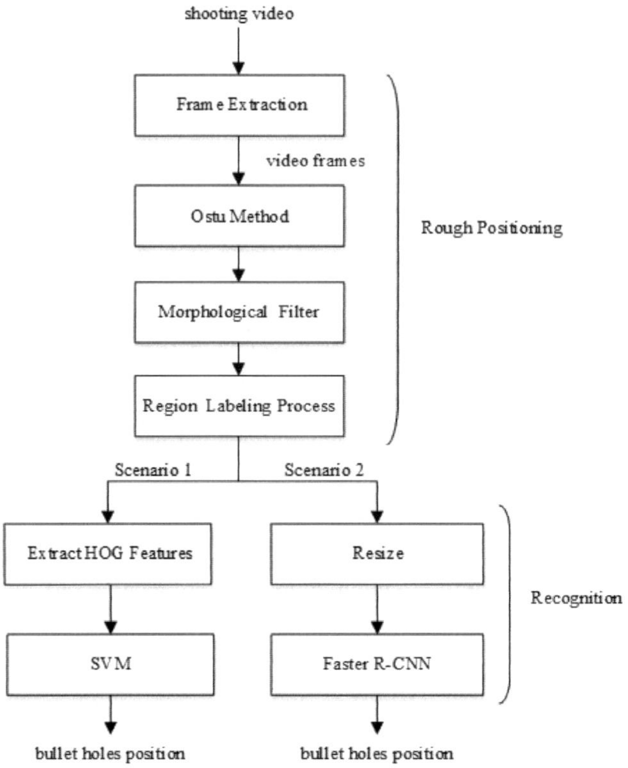

Figure 1. The process of bullet holes positioning.

2. Bullet holes rough positioning

2.1. Preprocessing
Before training, we need to preprocess the videos because of the noise interference. In this paper, the experimental data is the long-range shooting video for outdoor training with a variety of noise pollution. The target for long-range shooting is larger than normal target. As a consequence, the pixels of bullet holes in the video frame is much less. And the target is made of a soft cloth, which will be affected by wind and other external conditions readily.

First of all, We need to filter out environmental noise in order to get the position of bullet holes accurately, according to a prior knowledge about bullet holes. The median filter is used to filter out the ambient noise. After that, this paper uses Ostu method morphological filtering and region labeling process to complete the rough positioning of the bullet hole.

2.2. Rough positioning based on Ostu algorithm
Otsu method is based on the single order statistical characteristics of gray histogram, and makes maximizing the variance between classes of target regions and background as threshold selection criterion. In this paper, target regions in shooting video is regions of the bullet holes. Figure 2(a) shows results after the Otus algorithm.

But to shooting video frames, when background is a little complex and the ratio of bullet holes to background is very small, the Otsu method will falsely segment the less area of background into the results. So with considering the real-time processing requirement of the shooting videos, in this paper, Otsu method is only used for rough positioning of the actual bullet holes.

Then a part of areas that is too big or too small is removed by morphological filtering and region labeling process. Finally, record the suspected bullet holes as suspected images for recognition. Figure 2(b) draws suspicious samples on a frame image.

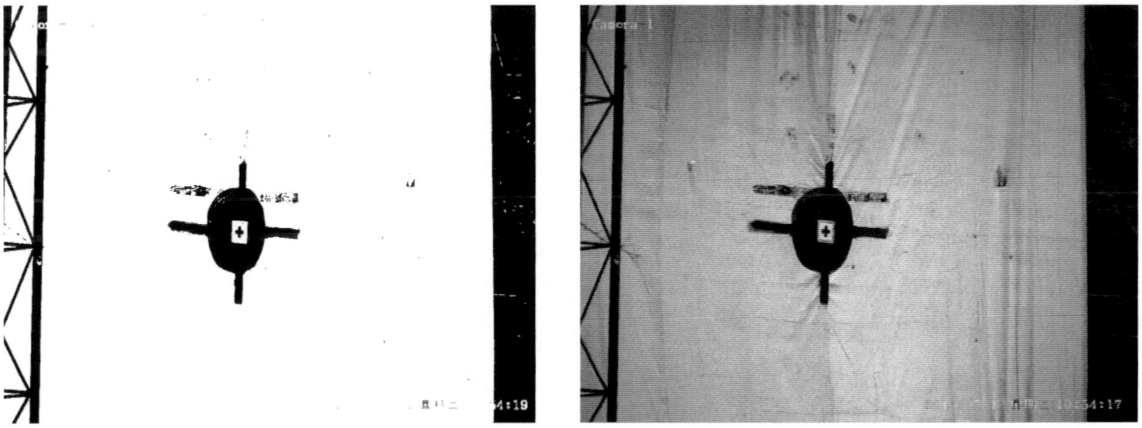

(a) Results after the Otus algorithm. (b) The frame image contains suspicious samples .

Figure 2. The result of bullet holes rough positioning

3. The recognition of bullet hole region based on shallow machine learning

3.1. Database construction

Training samples, in a way, impact on the algorithm's generalization ability. The experimental idea of this paper is to recognize actual bullet holes form the suspected images which obtained in the previous step. Considering in the actual shooting videos may have different ambient interference, we recorded videos in actual military shooting practice outdoor taken in natural backgrounds. And then make bullet holes rough positioning to videos that we recorded. After obtaining some suspected bullet holes areas by rough positioning, calculate the center of this connected areas, and cut the 24×24 image proposals on the original video frame with the center coordinates. Finally we obtain a total of 890 samples, of which 442 positive samples, while others are negative. The positive samples of database are shown in Figure 3(b), while Figure 3(a) are negative samples.

(a) The negative samples of database

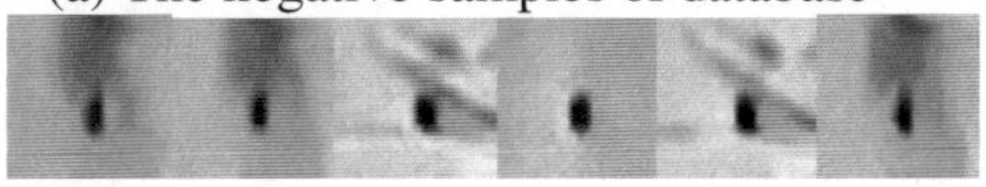

(b) The positive samples of database

Figure 3. The examples of database

3.2. The average gradient map

In order to ensure the distinguishing ability of HOG features, in the experiment, we draw the average gradient image and implements the visualization of the HOG features. When calculating HOG features, each sample produces a gradient image [4]. The gradient images of all positive samples from database are taken in average, and then normalized. The average gradient image of positive samples is shown in Figure 4. As average gradient image show, the characteristics of the bullet holes can be properly described by gradient.

(a) The average gradient image of dx (b) The average gradient image of dy

Figure 4. The average gradient image of positive samples

Visualizing HOG features, each direction corresponds to a bin [4], while the length is the gradient projection in this direction. Used to say that overlapping blocks lead to multiple updates of this sum, therefore we keep track how often a cell was updated to compute average gradient strengths. In the experiment, we change the size of cell and value of bin so as to obtain the best dimensional HOG features. Comparing the four pairs of in Figure 5, when size of cell choose 8×8 pixels and bin is 9, the characteristics of the positive and negative samples are the most distinguished. Figure 5(c) illustrates that this gradient projection are evenly distributed in all directions.

(a) when size of cell is 3×3 pixels and bin is 9, left one is the HOG visualization of positive samples, right one shows negative samples.

(b) when size of cell is 4×4 pixels and bin is 9, left one is the HOG visualization of positive samples, right one shows negative samples.

(c) when size of cell is 8×8 pixels and bin is 9, left one is the HOG visualization of positive samples, right one shows negative samples.

(d) when size of cell is 8×8 pixels and bin is 12, left one is the HOG visualization of positive samples, right one shows negative samples.

(e) when size of cell is 8×8 pixels and bin is 18, left one is the HOG visualization of positive samples, right one shows negative samples.

Figure 5. The the visualization of the HOG features

3.3. The recognition of bullet hole region based on HOG and SVM

As Figure 5 shown, negative samples are similar to bullet holes showing Intense changes in gradients, but differ in that this gradients are evenly distributed in all directions. HOG can properly describe the difference by histogram of oriented gradients. Histogram of oriented gradients contains measurements of direction and magnitude in each direction.

We propose HOG+SVM algorithms. In this experiment, HOG is used to extract gradients features from input images. Set block to 16×16 pixels and the stride to 4×4, so that each cell which is 8×8 resolution can be overlaped by some others. SVM is a typical binary classification algorithm[5], using it to distinguish between the true bullet holes and the others.

4. The recognition of books based on deep learning

Recent studies in image analysis have shown that the performance of convolutional neural networks (CNNs) is superior to that of shallow machine learning[6], owing to its capable of learning the essence of data. The network structure in deep learning is composed of the convolution layers of which the number is normally 3 or 5 or even more, pooling layers and the Fully connected layers. In this paper, we use the Faster R-CNN model that trained on the VOC2007 database[7] and then retrain it with the database we built ourself of bullet holes. Then images can be recognized and classified by the retrained model.

4.1. Training samples

In this experiment, training data choose the all positive samples from database which we built for bullet holes. But the experiments use images of 500×500 resolution for training, because the sample from the database is too small to locate bullet holes accurately. The samples of 24×24 resolution are amplified to images 500×500 resolution. Then label the images, putting them into network. The network will randomly select the size of crop_size × crop_size resolution as input.

4.2. Model Parameters and Experimental Process

Faster R-CNN is composed of two modules. The first module is a deep fully convolutional network called region proposal network(RPN), and the second module is the Fast R-CNN detector that uses the proposed regions. During the initializing phase, we set learning rate of 0.001 for 60 percent mini-batches, and 0.0001 for the others mini-batches on the database, the weight decay of 0.0005 and a momentum of 0.9. The weights are randomly initialized with zero-mean Gaussian function whose standard deviation 0.01. The Faster R-CNN model that adopted in this paper is shown in Figure 6.

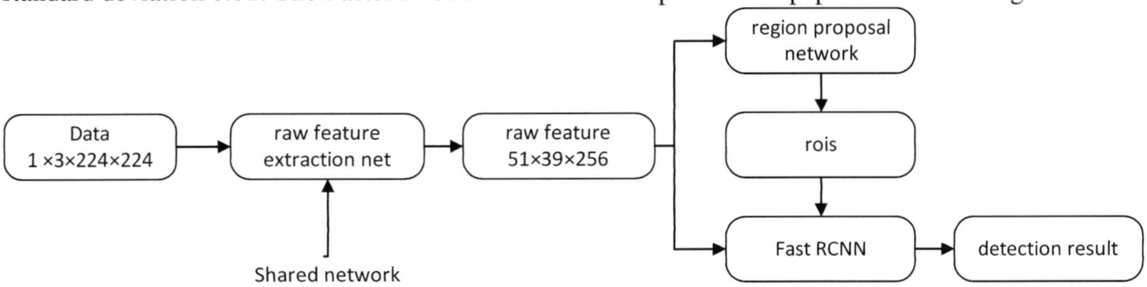

Figure 6. The Faster R-CNN model adopted in the experiment.

5. The comparison and analysis of two recognition methods

The testing samples using in the experiments are different from training database. We record a fresh video for Evaluating the performance of two algorithms. Select 160 frames from the vedio, and make a rough positioning.

This paper respectively adopts HOG+SVM and Faster R-CNN in deep learning algorithms, realizing the video image analysis of bullet holes. The recognition results are shown in Table 1.

Table 1. The comparison of two methods for the recognition of bullet holes

	Fp	Tp	Recall	Precision
HOG+SVM	30	2081	0.9294	0.9857
Faster R-CNN	358	2085	0.9312	0.8535

For the Recall, the deep learning method has shown a good performance, comparing the SVM. But for the Precision, performance of the shallow machine learning method is much better than it that Faster R-CNN shows. It means that features of bullet holes Faster R-CNN learned is not robust enough.

The reasons for this result may be that bullet holes are too small to study the inherent features, and the size of the training sample is inappropriate. To the samples of 24×24 resolutions, convolutional neural network can not get enough negative samples from whole frame images, while negative samples for SVM is different. For Faster R-CNN, to the small target such as bullet holes in this experiment, target image block can be shrunk to a point after several convolution operations. So the last layer of convolution layer has lost some of the features of bullet holes.

6. Conclusion

This paper studies two main methods which are shallow machine learning and deep learning algorithms. According the result of comparison experiments, analyze the recognition results from the aspects of Recall and Precision. In overall consideration, HOG+SVM algorithm is more suitable for the outdoor training of military shooting. The next step is to use information of adjacent frames in the video optimizing the algorithms.

Acknowledgments

The work on this paper was supported by the National Key Technology R&D Program (2015BAK22B02) and the Program (JXJYG1606) of Communication University of China.

References

[1] Zhou J 2014 Study of Military Automatic Target-scoring System Based on Machine Vision Technology

[2] Liu Q and Chen Y 2015 Research on Automatic Target-scoring Method Based on Video Image Analysis Computer Engineering

[3] Luo J, Zhang Z, Zeng G 2016 The study of multi-class classification based on support vector machine algorithm ITOEC **189**

[4] Dalal N and Triggs B 2005 Histograms of Oriented Gradients for Human Detection 1 **1.7w** p 886-893

[5] Zhu B, Yang L, Wu X and Guo T 2015 Automatic Recognition of Books Based on Machine Learning International Symposium on Computational & Busi.pp 74-78

[6] Tang F, Wang Z and Chen M 2005 The study of multi-class classification based on support vector machine algorithm Control and Decision Vol.20 No.7 pp 746-754

[7] Ren S, He K, Ross G and Sun J 2016 Faster R-CNN: Towards Real-Time Object Detection with Region Proposal Networks Microsoft Research arXiv:1506.01497v2 **200**

IOP Conf. Series: Materials Science and Engineering **261** (2017) 012021 doi:10.1088/1757-899X/261/1/012021

Exhibits Recognition System for Combining Online Services and Offline Services

He Ma[1, a], Jianbo Liu[1], Yuan Zhang[1] and Xiaoyu Wu[1]

[1]School of Information Engineering, Communication University of China, Beijing, China

E-mail: [a]mahe@cuc.edu.cn; [b]wuxiaoyu@cuc.edu.cn

Abstract. In order to achieve a more convenient and accurate digital museum navigation, we have developed a real–time and online-to-offline museum exhibits recognition system using image recognition method based on deep learning. In this paper, the client and server of the system are separated and connected through the HTTP. Firstly, by using the client app in the Android mobile phone, the user can take pictures and upload them to the server. Secondly, the features of the picture are extracted using the deep learning network in the server. With the help of the features, the pictures user uploaded are classified with a well-trained SVM. Finally, the classification results are sent to the client and the detailed exhibition's introduction corresponding to the classification results are shown in the client app. Experimental results demonstrate that the recognition accuracy is close to 100% and the computing time from the image uploading to the exhibit information show is less than 1S. By means of exhibition image recognition algorithm, our implemented exhibits recognition system can combine online detailed exhibition information to the user in the offline exhibition hall so as to achieve better digital navigation.

1. Introduction

With the awakening of people's national consciousness, they are increasingly aware of the importance and the charm of traditional culture.[1] More and more people have a strong interest in Chinese traditional culture.[2] Compared with other leisure venues, many people prefer to go to the Chinese Traditional Culture Museum. In the Chinese Traditional Culture Museum, the most attractive is the exhibits which is the representative of the Chinese Culture.

With the development of technology, people's needs seem to be unable to meet by manual interpretation. People want to understand the exhibits in a more convenient way. In order to meet the needs of people, a variety of digital navigation system is coming. Until now, there are two major kinds of digital navigation systems on the market:

Firstly, digital key. It is the most popular way used in the Chinese Traditional Museum. The client is a hand-held device, and the user can choose the goal explained by the speaker by pressing the key. Then the device will play the prepared description. This method does not require too much manual intervention, and it gives the users so complete freedom that the user can choose the exhibits he likes to be introduced. So it's very convenient.

Secondly, automatic induction. This method is based on the digital keys and add the auto play function. At present, the main way of induction is infrared automatic induction and wireless automatic induction. The kind of the induction requires the museum to equip the facilities. Such facilities not only require a certain amount of engineering construction but also may not be used due to the impact of the storage conditions of the exhibits. Therefore, the promotion of this approach has been greatly restricted.

The existing digital navigation mode has many disadvantages, such as the help of other equipment, the information provided cannot be saved and so on. Based on the shortcomings of the existing digital navigation, in this paper, we propose an online and offline interconnection system based on the recognition of the exhibits. Visitors can use the mobile phone to identify the kind of exhibits and know about the detailed information of this exhibit at anytime. The system uses the way of separating the client and server, which reduces the requirement of the equipment, and can not only reduce the recognition time, but also improve the efficiency of the recognition.

2. Implementation of the Exhibits Recognition

In this system, the most important steps is recognizing image. We use the features from the deep learning network and send them to the SVM to classify the pictures. And before setting up the network, we should prepare lots of pictures which are used to train the network.

2.1. Data Preparation

The exhibits mainly present in the system through the form of image, so the core technology of the system is image recognition. This system combines SVM with depth hierarchy features to achieve the identification of exhibits. In this system, we have to complete the model training offline, in order to make the model more robust and acquire better generalization ability. The training data is very important, we need to collect the field data to finish the training model.

We go to Jiading Museum, Shanghai Museum of art, Shanghai Museum and other museums. Finally, we choose the porcelain exhibit in the Jiading Museum as resource of the image data. In the process of acquiring the experimental data in the museum, we take the diversity of user equipment into consideration, so we use SLR camera, mobile phone, micro single, pad and other devices for data collection. In the process of collecting, we consider the complexity of real shooting situation, and do not have a single shot of exhibits: considering the diversity of the angle for taking the exhibits picture, we take photos of exhibits from different angles, covering the shooting angle the users use as much as possible; Taking into account the proportion and position that exhibits in the final picture. We take photos of exhibits from different locations, different distances. Taking different size exhibits photos, and making the location of the exhibits is not the same. Unfortunately, due to the special provisions of the museum, we cannot use our own light source to take photographs under different light conditions. Therefore, we get the photos under different lighting conditions by processing the image after taking photographs. In fact, the museum's light is relatively constant, the impact of light on the picture is relatively small. Therefore, the influence on the results of the identification is very small. The final selection of the types of porcelain and some of the pictures is shown in figure 1.

In order to make the system has the ability to reject, we also need to collect negative samples. The negative samples is other data which are not belong to sample categories. We take samples is divided into two parts, one part is the samples collected in Shanghai museum. Some of them is shown in figure 2. The other part is the pictures we are randomly selected from the data in Imagenet.

Figure 1. Some of the positive sample set.

Figure 2. Some of the negative sample set.

Due to the limited manpower, the photos taken in the museum is difficult to meet the experimental requirements. In order to make the training model have greater generalization ability and avoid overfitting of the convolutional neural network, we need to expand the data. The extension methods used in the system are the actual extension and the virtual expansion. The actual expansion refers to the expansion that the data will be saved to the local, while the virtual expansion will made images only for the network and will not save to the local. Therefore, we will make each class which have the 200 original images rotated, as well as the light change to expand the data. So each class of sample is expanded to 8000. And on the base of the actual expansion, the virtual expansion is mainly used with extended data in Caffe mode. The angle and light have random changes to complete the expansion of the data. Finally the proportion that data from negative samples and positive samples is almost 1:1.

2.2. The exhibition images classification based on CNN

After the data preparation, we can process the image with our own method. We use the CNN to extract the features of the image. Then send the features to the SVM. This method take the advantage of the CNN and SVM. The specific processing is shown in Figure 3.

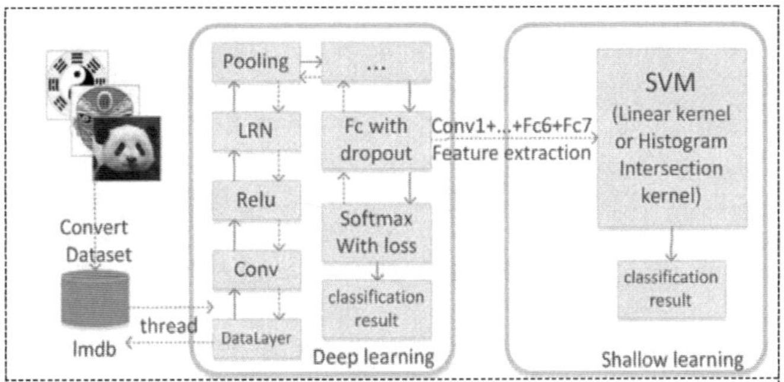

Figure 3. The process of the image processing.

2.2.1. Feature Extraction Using Deep Learning. With the introduction and development of the concept of deep learning, deep learning has played an important role in many areas. It is similar to the brain's deep cognitive process, which achieve multi-level data feature learning. Taking the image data used in the system as an example: Deep learning use the combination of the feature of the edge, the initial shape and other initial features to form a more abstract high-level representation. There are many models help us to apply the deep learning algorithms. Convolution Neural Network (CNN) is one of them, which is widely used in pattern recognition, image segmentation, and target detection.[3] And Alexnet is the classic model of image classification in CNN. Alexnet is the network made by Geoffrey and his student Alex in 2012. [4] Alexnet consists of five convolutions, two full layers and the last layer is the output layers. The server is mainly using the Caffe, to adjust the good mature model with a lot of data training, that is, fine-tuning.[5] Because the convolution network is to simulate the human brain's vision mechanism to identify different images. The more high-level features, the more abstract information we can get. The features are more able to represent the image. Thus, high-level features such as fc6, fc7 can better express the characteristics of the image. Therefore, we select fc6, fc7's output as the characteristics of the image.

The traditional network send feature information after fc7 directly into the softmax layer for classification.[6] But, whether the information has been abandoned is useless information or not? Considering the missing features, the server chooses to extract the features of the fc6 and fc7 layers. Each is characterized by a 4096 dimension, combining them into a large vector as a characteristic expression for various types of images. At the same time, we sent the features not into the softmax but the SVM classifier that we explains in the following chapters.

IOP Conf. Series: Materials Science and Engineering **261** (2017) 012021 doi:10.1088/1757-899X/261/1/012021

2.2.2. SVM. The nature of the identification problem is actually a classification problem, the last layer of the traditional neural network is connected to the softmax. The softmax regression model using the gradient descent to update the parameters. Because of the existence of this process, the speed of convergence will slow down. At the same time, because of the interference from the interference group data, the classification accuracy will decline. However, the traditional SVM classifier does not need to calculate the cost function. So the speed of convergence will be better than the softmax. SVM classifier has another great advantage that it has good performance on anti-noise.[7] Therefore, the server does not use the traditional network to classify the feature. But send the feature that the network extracts to the traditional SVM classifier for classification.

The main idea of SVM is increase dimension of input space to ensure it is Linearly Separable in high dimension space by using the method of nonlinear mapping. SVM uses the expansion theorem of the kernel function to avoid the curse of dimensionality. Different kernel functions are selected to generate the different SVM. The kernel functions that we use is the histogram intersection kernel function.

SVM classifier is mainly used to solve bipartition. In the real world most of the data is multi-class data, so the simple SVM need further expansion, so that it can solve the multi-class classification. There are many methods can realize the expansion of SVM. The system uses the one to one , that is, design a classifier between any two samples, so the K types need to design K (k-1) /2 classifier. When an unknown sample is classified, the class with the highest number of votes is the sample.

The use and construction of SVM classifier consists of two parts: training and recognition

Firstly, offline training process: the feature that extracted by the deep learning network will be sent to the SVM for training. Finally, we will get a trained SVM classifier.

Secondly, online identification process: similarly, after convolution neural network extract the feature of the test image's fc6 and fc7 layer, the features will be sent into the SVM classification has been trained. And SVM will give the final classification results.

3. Design and Implementation of the system

In order to meet the requirements that we can identify more kinds of exhibits with less time and less resource. In our system, we use the communication between the client and the server to realize the identification of the exhibits. The photo of the exhibit will be uploaded to the server through the Internet. The server return the identification result to the client after identification. The client show the result with the text.

3.1. The Client

The client is based on Android 6.0.1 or above systems. Like the ordinary app, you can install software after downloading the installer. The client's interface is easy to understand and the operation is easy. From the main interface of the client we can clearly see all the function of the software. The software can not only recognizes the captured picture, but also realize the recognition of two-dimensional code. And with the help of communication of the Baidu Server, the function of text recognition can be realized.

As the most important function of the software, the interactive logic of the real-time image capture and recognition is very simple. When users decide to take a photo, they just need to press the camera button in the bottom left corner. The moment they press the camera button, the client will set up connections with the server and send the picture data to the server. After the server recognizes the picture, the number of the class that the picture belongs to will be returned to the client. After the client accepts the number, the name of the porcelain corresponding to the number and its related description are displayed on the interface.

At the same time, we add some auxiliary functionality to help the users use the app more convenient. Clicking the library button in the bottom left corner, the user can select photos from the gallery. And clicking the record button in the bottom right corner, it will show the picture has been identified. The users can delete records and view the recognition results of the picture has been identified.

3.2. Communication between Client and Server

After the completion of the exhibits image's collection in the client, the image information must be uploaded to the server. And the result can be displayed after processing by the server. Therefore, the communication between the client and the server is an important step in the implementation of the system. There are many ways that can realize the communication between the client and the server.

Finally we choose the Http protocol for transmission. The Http protocol is the object oriented application layer's protocol. Because only the request method and path transmission are needed to, the Http server program is very small. Therefore, the communication speed is very fast. Http protocol is flexible enough to allow the transmission of any type of data object. The process using Http protocol to achieve the connection between the client and the server mainly includes the following steps:

Firstly, the client sends the request to the server. Because in our system, the client needs to upload the image data, and the server needs to return the information, we choose post (Post ask the serve receive the data attached to the request). We attach the data in the form to the request.

Secondly, server response the request and send the result to client. The server will process the image data that sent by the client. The server will get the number of the category in the image after recognition. The server will attach the results of recognition to the response to the request of the client. When the client receives the message sent by the server, the connection ends. The Http transport process is complete.

4. Experimental Results

In this experiment, we use a PC as a server, and huawei honor 7i smart-phone as the client, whose configuration is 8 core 1.5 GHz processor, memory is 3.0 GB and mobile phone system is android 6.0.1 system. To achieve the communication between server and client, we use the campus network. In a relatively stable environment, we do the test, using the phone's inbuilt speed test tool to test the speed of the network. Of course, the background software was clean. The average upload speed is about 63.7 KB/S, and the average download speed is about 371 KB/S.

Due to the limited conditions, we choose to take photos of the picture taken from a museum from different perspectives. We randomly selected five photos of each porcelain to do this experiment. And the number of taken photos was 100. Part of the picture are shown in figure 4.

In the end, the recognition accuracy is 100%.Because our system is real-time, recognition time is what we concern most. In the experiment, the average recognition time for 599 ms. Among them, the shortest recognition time is 329 ms. All recognition time within 1000 ms, the longest recognition time is 975 ms. The distribution of recognition time is shown in figure 5.

Figure 4. Some of the pictures which have been used in the experiments.

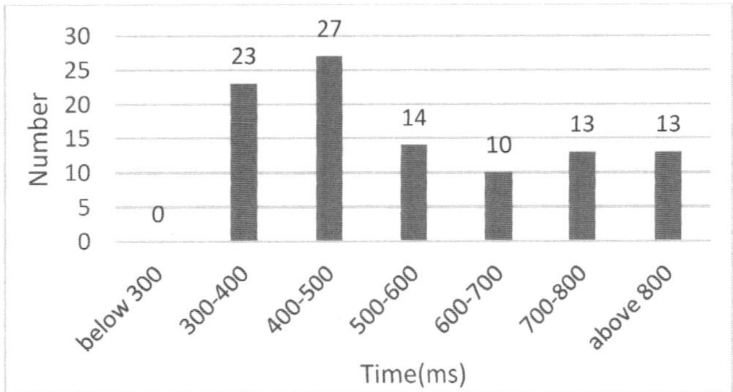

Figure 5. The proportion of the recognition time distribution

5. Summary

In this paper, we introduce the exhibits recognition system for combining online services and offline services. The client of this system is the phone. The visitors can use the phone to take the pictures of the exhibits. Then the data will be sent to the server. The server is used to recognize the image. We use the deep learning network to extract the features of the image and send features to the SVM to classify. What is worth mentioning is, in this system, in order to make the model more robust and acquire better generalization ability, we have to complete the model training offline. After recognition, the result will be returned to the client and will be shown in the phone.

To reach a higher speed and use less time to complete the communication between the client and the server, we use the HTTP transport. Finally, in our experiment, we find the recognition accuracy is 100% and the time of one process is very short.

The online and offline interconnection system based on exhibits identification proposed in this paper has the following advantages compared with the existing navigation system:

Firstly, the server uses a combination of deep learning and SVM classifier, with a higher recognition rate.

Secondly, the use of the way of separating the client and server makes the requirements of the client greatly reduced.

Finally, using a smart phone as a client, more simple and convenient.

Experimental results show that the system is feasible and practical, in the era of smart phones, the navigation based on smart phone will bring the difference to the museum's navigation system.

Acknowledgments

The work on this paper was supported by the National Key Technology R&D Program (2015BAK22B00) and the Program (JXJYG1606) of Communication University of China.

References

[1] Tan X, Wu X, Yang C, Shen Y 2016 Proceedings of the 20165th IEEE International Conference on Image, Vision and Computing Chinese Traditional Visual Cultural Symbols Recognition based on SPM Muti-feature Extraction C

[2] Tan X, Wu X, Yang C 2015 Proceedings of the 2015.8th International Symposium on Computational Intelligence and Design -Volume 02. IEEE Computer Society 2015.304 Visual Cultural Symbol Recognition Based on Muti-feature Extracting C

[3] Dahl J V, Koch K C, Kleinhans E, et al 2010 Convolutional networks and applications in vision J 14(5):253-256 345

[4] Krizhevsky A, Sutskever I, Hinton G E 2012 Advances in Neural Information Processing Systems ImageNet Classification with Deep Convolutional Neural Networks J 25(2) 7442

[5] Jia Y 2013 http://caffe.berkeleyvision.org Caffe: An Open Source Convolutional Architecture for Fast Feature Embedding 2435

[6] Deng J, Dong W, Socher R, et al. 2009 ImageNet: A large-scale hierarchical image database C 248-255 3212

[7] Zhang X 2000 Acta Automatica Sinical, Vol.26, No.1, pp 32-42 1481

IOP Conf. Series: Materials Science and Engineering **261** (2017) 012022 doi:10.1088/1757-899X/261/1/012022

Mapping High Dimensional Sparse Customer Requirements into Product Configurations

Yao Jiao[1], Yu Yang[1,*] and Hongshan Zhang[1]

[1]Department of Industrial Engineering, College of Mechanical Engineering, Chongqing University, Chongqing, China

[1,*]yuyang@cqu.edu.cn

Abstract. Mapping customer requirements into product configurations is a crucial step for product design, while, customers express their needs ambiguously and locally due to the lack of domain knowledge. Thus the data mining process of customer requirements might result in fragmental information with high dimensional sparsity, leading the mapping procedure risk uncertainty and complexity. The Expert Judgment is widely applied against that background since there is no formal requirements for systematic or structural data. However, there are concerns on the repeatability and bias for Expert Judgment. In this study, an integrated method by adjusted Local Linear Embedding (LLE) and Naïve Bayes (NB) classifier is proposed to map high dimensional sparse customer requirements to product configurations. The integrated method adjusts classical LLE to preprocess high dimensional sparse dataset to satisfy the prerequisite of NB for classifying different customer requirements to corresponding product configurations. Compared with Expert Judgment, the adjusted LLE with NB performs much better in a real-world Tablet PC design case both in accuracy and robustness.

1. Introduction

Customers express their requirements on product attributes ambiguously and locally due to the lack of expertise [1]. The ambiguous requirements urge designers to develop precise configurators for mapping uncertain information to accurate product specifications. The local requirements form various differentiation on disparate product attributes corresponding to customers' finite and variant domain knowledge. Therefore, mining data of customer requirements from various mediums or direct expressions might lead to fragmental data with high dimensional sparsity, burden the process for mapping customer requirements to product configurations with uncertainty and complexity [2].

Popular methods for mapping customer needs to product configurations heavily rely on the explicit, systematic and structural requirements data [3], due to the lack of expertise, customers may unpleasant and stressful to raise their needs specifically and terminologically [1]. Meanwhile, collecting data for reducing sparsity is time consuming and costly, it is difficult to guarantee the data collected is accurate and consistent [4]. These obstacles make the uncertain, high dimensional sparse dataset of customer requirements arise from many real-world design issues.

In face of limited data availability and fuzzy information, the expert judgement is widely used in practice, as it has no such requirements for data compared with other popular methods for mapping procedures. However, expert judgement raises concern on its repeatability and bias [5]. Some researches emphasize the probabilistic classifiers of machine learning for they are suitable to quantify uncertain information, and robust in diversity [6-7]. Such probabilistic classifiers are inductive learning techniques, the training datasets from which the classifiers learn the regulations should be under basic prerequisites. Therefore, to apply probabilistic classifiers in product design issues, preprocesses for high dimensional sparse dataset are necessary.

In this study, we apply Naïve Bayes (NB) classifier for mapping customer requirements to product configurations. NB makes use of requirement flexibility, and calculate probability of relevance for all product configurations. This merit makes NB capable and efficient in fuzzy reasoning. Another merit is it converges fast in learning stage, thus NB is suitable for small training dataset which is usual in design issues. However, NB is built on the prerequisite of conditional independent assumption. High dimensional dataset contains so many intertwined relationships and dependencies between attributes, which may cause NB inefficient and fallible, even results in the curse of dimensionality [8].

Therefore, before implementing NB for mapping reasoning, this study uses a dimensionality reduction method, local linear embedding (LLE), to preprocess the high dimensional requirement dataset. Considering the sparsity, some adjustments are applied based on basic LLE to improve its capability for processing sparsity. In addition, this study uses the data from a real-world design case to compare the NB integrated with adjusted LLE to expert judgement in the performance of mapping customer requirements to product configurations.

The rest of this paper is organized as follows. In the section of related works, we review methods for mapping customer needs to product configurations. In the next section, detailed methodology of this study is given out. A real-world Tablet PC design case and its data is shown in the section of data and case, then followed by the results and discussions. At last, we conclude the paper with contribution and future works.

2. Related work

Several approaches are available for mapping customer requirements into appropriate product configuration plans, e.g., configurator design, Kansei engineering, and affective computing [3]. In configurator design process, customers are required to specify product attributes with one product one by one for resulting the desire product [9]. Configurators can be divided into 3 categories, as rule based, model based and case based. Rule based configurators derive configurations in a forward manner [1]. It is firstly developed and applied in by Digital Equipment Corporation with the name called R1. R1 succeeds in application greatly, and accelerates other configurators based on rules, e.g., SICONFEX, MMC-Kon, BLADES, and MICON [3,9]. However, due to the lack of separation of domain knowledge and control system, such rule based systems often suffer from maintainability issues or side effects from rule changing, and fail to accomplish complex design works [6]. Model based configurators take configuration tasks as constraint satisfaction problems (CSP) [10]. Model based systems have advantages of better separation of domain knowledge used, and better robust for complex design tasks [7]. Extensive research of such model based configurators accompanied by the deep-going study of CSP, e.g., the Dynamic CSP contains activity constrains, the Generic CSP for handling constrains in intuitive way, the Composite CSP, and the Mixed and conditional CSP [1,3]. Case based configurators store previous analogous configuration tasks and domain knowledge, then the new configurations are retrieved from the storage [11]. The basic issue in case based systems is how to find similar solved problems and retrieve the best configuration adapting to the new requirements. In summary, challenges for configurators exist in several areas, e.g., knowledge compilation, conflict detection and diagnosis of divide & conquer algorithms, and interactive settings [12]. Besides, configurator design burden customers express their needs in specific parameter domain rather than customer requirement domain [1]. This makes configurator design difficult for mapping ambiguous requirements to specific detail design parameters.

Kansei engineering is introduced by Nagamachi as a translating technology of a customer's feeling of the product to the design parameters. Kansei in Japanese means the psychological feeling and image regarding of customers to the new product, thus Kansei engineering is considered as a methodology within Affective Engineering [13]. Kansei engineering starts with the choice of domain to describe the overall idea behind an assembly of products for further evaluation. The second step is the selection of kansei words from spanning the Semantic Space which is the collection of words that describe the domain, e.g., emotions, moods, and impressions. Then, similar to the spanning of Semantic Space, the third step is to span the space of product properties. In the next, synthesis step, Semantic Space and the space of product properties are linked together, to form the database that quantifies the connections between those two spaces. Kansei engineering is very helpful for customers to select the most satisfied product to his/her kansei [14]. However, Semantic Space demands explicit expressions of customer feelings and psychological activities to form the kansei words.

Making customers involve in design process is commonly adopted for processing ambiguous customer requirements [15]. By making customers immerse into design process, designers can acquire customer needs directly from their knowledge, behaviors, and reflections to products, to avoid the false information from traditional market research methods of survey and interview, e.g., focus group, quality function deployment, conjoint analysis, etc. [16]. The lead user design and empathic design are two typical methods for inviting customers to design process. Lead user design encourage those customers, who push a product to its limits and experience needs prior to commons, to distill the needs that satisfy themselves significantly by customizing. Though lead user design is effective for open innovation, those lead users are difficult to find, and may suggest configuration plans for niche products or higher-end products, rather than mainstream products [17]. Empathic design elicits customers' unarticulated requirements by observing customers using products in everyday applications and usage contexts [18]. This foundation makes empathetic design resultful in identifying latent customer needs. However, empathic design depends heavily on designer's power of observation and innovation, and can hardly implement in complex products for the consideration of simulation [17].

Expert judgement is a widely practiced technique for product design in the uncertain and sparse dataset settings, as it has no formal requirement for systematic data which is quite common in design issues [4]. Due to the dependence upon expert knowledge, expertise and experience, repeatability and bias are concerns raised for expert judgement [19]. These concerns demand for the validation to the accuracy and informativeness of expert judgement applications [20]. Most cases of expert judgement reduce to cross validation, whereby experts learn experience from training set, and make judgements on the test set. In addition, taking at least 80% of the entire dataset for training is suggested [21].

3. Methodology

Expert Judgment is a commonly applied method in high dimensional sparse dataset settings, since there is no formal requirement for systematic and structural data. The Figure 1 shows the overview of mapping procedure by using Expert Judgment. In this study, we develop an integrated method by adjusted LLE and NB for mapping unarticulated high dimensional sparse customer requirements to product configurations, where the adjusted LLE preprocess the high dimensional sparse dataset to satisfy the prerequisite of NB for classifying. The overview of this integrated method is shown in Figure 2. Moreover, we compare the proposed method to Expert Judgment in a real-world design case. The next subsections introduce (a) the adjustments for classical LLE to reduce dimensionality with sparsity, (b) the classification process by using NB, and (c) the design for comparison with proposed method and Expert Judgment.

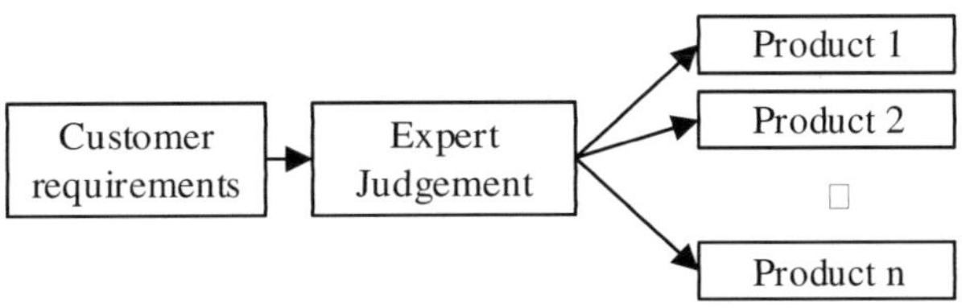

Figure 1. Mapping customer requirements based on Expert Judgement

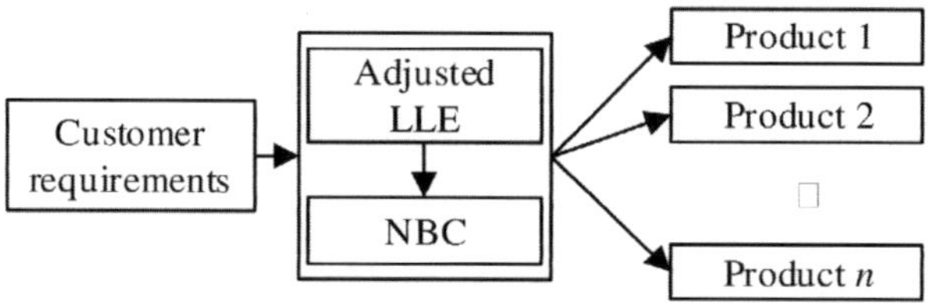

Figure 2. Mapping customer requirements based on adjusted LLE and NB

3.1. Adjusted LLE for sparse data dimensionality reduction

LLE is nonlinear dimensionality reduction method based on manifold learning, and introduced by [22]. For detailed interpretation of LLE, see [22]. LLE has many advantages over other manifold learning methods for dimensionality reduction, because it is able to learn global structure of nonlinear manifolds, which makes it very suitable for pragmatic application in real-world settings. However, it is limit when dealing with sparse data, due to it attempts to retain only local properties of the data [23].

In this study, we adjust LLE by expanding the initial local neighborhood to against the sparsity of customer requirements. The adjusted LLE is described as follows.

For a given high dimensional sparse dataset $X = [x_1, x_2, \cdots, x_N]$, $x_i \in R^D$, D is the dimensionality, suppose its low dimensional representation dataset $Y = [y_1, y_2, \cdots y_N]$, $y_i \in R^d$, d is the dimensionality, where $d \ll D$. For each x_i, finds k nearest neighbors. For adjusted LLE, the expanded neighborhood of x_i is the combined neighborhood of nearest neighbors of the initial k nearest neighbors of x_i.

Let $N_k(x_i) = \{x_j\}$ is the set of k nearest neighbors of x_i, $i \neq j$, $j = 1, 2, \cdots, k$, the expanded area is $N_k^E(x_i) = \{N_k(x_i), N_k(x_j)\}$. We apply the tangent space discriminant (TSD) to select and retain the sample datapoints. The Eigen-decomposition process of TSD requires at least $m+1$ local sample datapoints, where m is the intrinsic dimensionality of X, in addition, the number of expanded nearest datapoints $k^* = k^2 + k$. Select the m first principal eigenvectors corresponding to smallest k^* nonzero eigenvalues of every sample datapoint, where all the eigenvectors are subject to $\lambda_1 \geq \lambda_2 \geq \cdots \geq \lambda_m \geq \cdots \geq \lambda_{k^*}$, to form the columns of matrix H, and H_i denotes the tangent direction of x_i. Then, an expanded local sample datapoint can be retained if it satisfies Eq. (1):

$$\left|\left(I - H_i H_i^T\right)\left(x_i - x_j^*\right)\right|_2 < T^*(x_i), \tag{1}$$

where I is the identity matrix, $x_j^* \in \{N_k(x_j)\}$, $j = 1, 2, \cdots, k$, $T^*(x_i) = \left(1/k^*\right)^{1/m} \left|x_i - x_j^*\right|$.

Then the expanded local neighborhood $N_k^{E*}(x_i)$ is formed by all the retained sample datapoints which satisfy Eq. (1). Calculate a sparse matrix of local predictive weights w_{ij} that best reconstruct every datapoint x_i from its k nearest neighbors as:

$$\min \varepsilon(W) = \sum_{i=1}^{N} \left\| \sum_{j=1}^{k} w_{ij}\left(x_i - x_j\right) \right\|^2 = \sum_{i=1}^{N} \left\|\left(x_i - x_j\right) w_i\right\|^2 = \sum_{i=1}^{N} \left(w_i\right)^T Z_i w_i, \ s.t. \ \sum_{j=1}^{k} w_{ij} = 1, \tag{2}$$

where $Z_i = \left(x_i - x_j\right)\left(x_i - x_j\right)^T$. Based on Lagrangian multiplier methods (LMM), $\partial L / \partial w_i = 2Z_i w_i + \lambda^* I_N = 0$, we can derive $w_i = Z_i^{-1} \cdot I_k \left(I_k^T Z_i^{-1} I_k\right)^{-1}$.

To retain the weight of $N_k^{E*}(x_i)$, the cost function should be minimized as:

$$\min \Phi\left(Y_{N_k^{E*}(x_i)}\right) = \sum_{p \in N_k^{E*}(x_i)} \left\| y_p - \sum_{j=1}^{k} w_{pj} y_j \right\|^2. \tag{3}$$

Therefore, Y can be computed as:

$$\min \Phi(Y)=\sum_{i=1}^{N}\sum_{p\in N_k^{F*}(x_i)}\left\|y_p-\sum_{j=1}^{k}w_{pj}y_j\right\|^2=\sum_{i=1}^{N}\sum_{p\in N_k^{F*}(x_i)}\left\|Y_i(I_p-w_p)\right\|^2=\sum_{i=1}^{N}Tr\left(Y_iM_iY_i^T\right), \quad (4)$$

where I_p is the pth column of $I, M_i=(I-W_i)(I-W_i)^T$.

Suppose $S=[S_1,S_2,\cdots S_N]$, $Y_i=YS_i$, $i=1,2,\cdots,N$, where $S_{ij}=1\ (i=j)$ or $S_{ij}=0$, and $M=diag(M_1,M_2,\cdots,M_N)$, $B=SMS^T$, thus Eq. (4) can be rewritten as:

$$\min \Phi(Y)=\sum_{i=1}^{N}Tr\left(Y_iM_iY_i^T\right)=\sum_{i=1}^{N}Tr\left(YBY^T\right),\ s.t.\ YY^T=I. \quad (5)$$

Based on LMM, $BY^T=\lambda Y^T$, Y is the eigenvectors corresponding to the d smallest nonzero eigenvalues of B. Therefore, the representation dataset $Y=[y_1,y_2,\cdots y_N]$, $y_i\in R^d$ replaces X of original customer requirement to be the input of NB process.

3.2.NB for classifying process

In this NB process, dataset Y is the input for replacing initial customer requirements dataset X, to meet the prerequisite of NB. To map customer requirements to the optimal configuration plan, $P\left(\overline{U}\mid y_1,y_2,\ldots,y_N\right)$ is needed, where $\overline{U}$ denotes a configuration plan, in addition, then:

$$P\left(\overline{U}\mid y_1,y_2,\ldots,y_N\right)=P\left(\overline{U}\right)P\left(y_1,y_2,\ldots,y_N\mid\overline{U}\right)\left(P\left(y_1,y_2,\ldots,y_N\right)\right)^{-1}\propto P\left(\overline{U}\right)\cdot\prod_{k=1}^{N}P\left(y_k\mid\overline{U}\right). \quad (6)$$

Let u_i denotes the ith product configuration, [24] estimate the probabilities of $P\left(\overline{U}\right)$ and $P\left(y_k\mid\overline{U}\right)$ by applying maximum likelihood estimation in the learning stage for NB, as:

$$P\left(\overline{U}=u_i\right)=\left|u_i\right|\left(\sum_i\left|u_i\right|\right)^{-1},\ \text{and}\ P\left(y_k\mid\overline{U}=u_i\right)=\left|y_k\cap u_i\right|\left(\left|u_i\right|\right)^{-1}, \quad (7)$$

where $|u_i|$ denotes the number of product configuration u_i, and $|y_k\cap u_i|$ is the number of cases of the choosing of u_i. The learning stage can be applied by using existing customer requirements representation dataset Y and corresponding selected configuration plans.

The NB is trained to select the configuration plan with highest probability, as:

$$\overline{U}^*=\arg\max_U P\left(\overline{U}\mid y_1,y_2,\ldots,y_N\right)=\arg\max_U P\left(\overline{U}\right)\cdot\prod_{k=1}^{N}P\left(y_k\mid\overline{U}\right). \quad (8)$$

3.3. Cross validation and performance measurements

Ten-fold cross validation is applied in this study to verify the effectiveness of the methods proposed over Expert Judgment. For a given high dimensional sparse customer requirements dataset, nine tenths of the data are randomly selected as the training set, and the remained one tenth act as the testing data. Such cross validation procedure is carried out for ten rounds. In each round, the training data is learned by experts, simultaneously, preprocessed by adjusted LLE and learned for NB, then the performances of Expert Judgment and adjusted LLE with NB are compared under two performance measurements for the testing data, they are: (*a*) The accuracy rate, and (*b*) the F-measure. In the following content, the measurements for accuracy and F-measure are given out.

Let π denotes the overall accuracy of all classes, m denote the number of class, TC_i denotes the number of classifications that correctly assigned to class i, FC_i denotes the number of classifications that not belong to class i, then, the overall accuracy rate is given by:

$$\pi=\left(\sum_{i=1}^{m}(TC_i+FC_i)\right)^{-1}\sum_{i=1}^{m}TC_i. \quad (9)$$

However, accuracy rate omits the false negative rate when measures the performance [1]. Thus, the F-measure is applied to overcome this issue. F-measure is originally defined for binary classification case. This study adopts macro-average F-measure proposed by [25] for multiple classification cases. F-measure ranges from 0 to 1, with higher value indicates better classification quality.

Let ρ_i denotes the recall rate of class i, FI_i denotes the number of individuals which are not assigned to class i while they actually belong to. Then, the recall rate is given by:

$$\rho_i = TC_i \cdot (TC_i + FI_i)^{-1} \tag{10}$$

Let π_i denotes the accuracy for class i, The macro-average F-measure F_{macro} is calculated as:

$$F_{macro} = m^{-1} \sum_{i=1}^{m} \left(2\rho_i \pi_i \middle/ \rho_i + \pi \right) \tag{11}$$

4. Case and data

Data from a real-world customized tablet PC design project is used to testify the effectiveness of the method proposed in this study. The design project is supported by a consumer electronics company in south China. There are 126 attributes to be specified by customers which corresponding to the requirements. Number of options alternatives for attributes ranges from 3 to 24, thus the customer requirements construct a 126×24 matrix. Part of the attributes and their options are listed in Table 1.

Table 1. Part of the attributes and corresponding options alternatives.

Attribute	Attribute Options			
Operation system	Android	Windows	Android +windows	
Screen size	7.9 inch	8.4 inch	9.7 inch	10.1 inch
CPU cores	2 cores	4 cores	6 cores	8 cores
CPU brand	Intel	Nvidia	MTK	Qualcomm
Color	Black	Gray	Yellow/gold	White
Storage volume	16G	32G	64G	128G
Aspect ratio	3:2	4:3	16:9	16:10
HDMI interface	None	Micro HDMI	Standard HDMI	

1030 customers were interviewed in a survey in two stages. At the first stage, customers were required to express their requirements to the tablet PC, and specify their expression under the assistance of designers. After that, they were shown a set of 24 different tablet PCs, and were required to make selection based on their requirements and preferences. Thus, there are 24 classes for classifying in this case. The customer requirements dataset and corresponding final selection will be used as the training and testing data for the ten-fold cross validation and performance measurements.

Expert Judgments for each round is decided by an expert team with eight experts. The expert team contains four senior designers in electronic products, two professors mainly research on consumer products design, and two executive editors from magazines focus on electronics products.

Sparsity is defined as the proportion of non-zero elements in a matrix [26]. The useful elements take account for 27.15% of the requirements matrix, but considering zero elements in the useful portion as customers do not take all the provided options, the sparsity is as high as 0.8456.

5. Results and discussion

Before comparing the performance of adjusted LLE with NB to expert judgement, this study has conducted a comparison of the number of extracted features (Num. of ET) and the computing time for classic LLE and adjusted LLE under the case situation. The results are shown in the Table 2.

Table 2. Comparison of LLE and adjusted LLE for extracting features.

Algorithm	Num. of ET	Computing time
LLE	32	2.742
Adjusted LLE	6	1.411

The results in Table 2 indicate that due to the expansion of initial nearest neighbors, the adjusted LLE can induct rules at the very beginning for reduction. While, the classic LLE costed more computational time because it fails to learn inductions in the first trials, the re-trials for finding enough initial datapoints to construct inductive rules are time consuming. Also, the number of features extracted by LLE is more than adjusted LLE, which implies the poor performance of LLE for reducing high dimensional spare data over the adjusted LLE proposed in this study.

Table 3. The accuracy rate and F-measure for two methods in every round.

Round	Adjusted LLE with NB		Expert Judgment	
	Accuracy rate (%)	F-measure (%)	Accuracy rate (%)	F-measure (%)
1	81.55	84.17	49.51	55.28
2	79.61	82.50	52.43	57.08
3	82.52	85.00	45.63	52.92
4	83.50	85.83	53.40	51.39
5	81.55	84.17	47.57	53.75
6	79.61	83.33	50.49	56.94
7	80.58	83.33	52.43	57.78
8	81.55	85.00	41.75	48.47
9	82.52	85.83	43.69	50.69
10	82.52	84.17	48.54	55.42

Table 4. Statistical analysis for accuracy and F-measure.

Measurement	Mean	Std. deviation	Kurtosis
Adjusted LLE with NB			
Accuracy rate	0.816	-0.748	-0.748
F-measure	0.843	-0.751	-0.751
Expert Judgment			
Accuracy rate	0.485	-0.805	-0.805
F-measure	0.540	-0.806	-0.806

Table 3 shows the results of accuracy and F-measure for each round of the adjusted LLE with NB and Expert Judgement based on the case data. Table 4 shows statistical analysis based on the results data shown in Table 3. All the results indicate that adjusted LLE with NB proposed in this study performance better than Expert Judgment in mapping high dimensional sparse customer requirements to product configurations in our research.

In each round, adjusted LLE with NB outputs higher accurate result (see 'Accuracy rate' in Table 3) and higher classification quality (see 'F-measure' in Table 3). Statistical analysis for the results indicates that adjusted LLE with NB are more robust than Expert Judgment. This verdict can be revealed from two aspects, (*a*) results of accuracy and F-measure of adjusted

IOP Conf. Series: Materials Science and Engineering **261** (2017) 012022 doi:10.1088/1757-899X/261/1/012022

LLE and NB are less dispersive (see 'Std. deviation' in Table 4), (*b*) Expert Judgment is more possible to generate outliers in classification process (see 'Kurtosis' in Table 4).

Expert Judgment is easy to implement and is widely used, but this study raises three major issues for implementing Expert Judgment in high dimensional sparse customer requirements settings of product design as follows:

- Experts selection. Experience of expert impact on the performance of Expert Judgment greatly. Thus, the first obstacle for Expert Judgment is the selection for composing members of an expert team, to reduce impact of differentiation in expertise. This study selects eight experts with ample expertise, however, the results of Expert Judgment are more dispersive than machine learning method.
- Low accuracy. The average accuracy rate of Expert Judgment is below 0.5 for the high dimensional sparse customer requirements in this study. It challenges the reliability of exert judgment for processing unsystematic and nonstructural dataset with sparsity. The low accuracy of Expert Judgment in this study reminds the necessity of analysis of suitability for its application in real-world design issues.
- High cost. The expert team costs at least three days for each round in learning from training dataset and determining classes for testing dataset, and the Expert Judgment in this study lasts more than one month. Though Expert Judgment is easy to implement, in front of high dimensional sparse customer requirements, it is costly, time consuming, and disadvantageous to make advantages for product design.

To the contrary, adjusted LLE with NB overcomes the major issues of Expert Judgment, and has several advantages. Firstly, the outputs of adjusted LLE with NB are of high accuracy and robust, make it reliable to process high dimensional sparse customer requirements. Secondly, adjusted LLE and NB is not difficult to implement pragmatically. Moreover, experts are not necessary for computing and reasoning in the procedures of adjusted LLE and NB, which saves time and cost due to human factors.

6. Conclusion

In this study, we propose an integrated method by adjusted LLE and NB for mapping high dimensional sparse customer requirements to product design configurations. And we compare this method with widely used expert judgement in coping with a real-world Tablet PC design case with unarticulated high dimensional sparse customer requirements. The results indicate this method outperforms expert judgement greatly both in accuracy and robustness. The contribution of the paper is summarized as follows:

- We apply NB to map customer requirements to product configurations. Compare with other methods, such as configurator design, Kansei engineering, etc., NB is a probabilistic classifier that more suitable to quantify and process the ambiguous and unarticulated nature of customer requirements. Thus customers are not burdened to express their needs specifically.
- The classical LLE is efficient in dimensionality reduction. In this study, we adjust the classical LLE by expanding initial local neighborhood to guarantee its effectiveness for dealing with sparse dataset, and make raw dataset satisfy the prerequisites of NB. The integration of adjusted LLE and NB performs better in coping with real-world design case than the widely used Expert Judgment under high dimensional sparse customer requirements settings, which is common in design issues with big data.
- The proposed method, adjusted LLE with NB, performs robustly, and no need for expertise or experience from experts to conduct computing and reasoning procedures, as LLE and NB are methods based on machine learning that learn rules by induction. These characteristics make the method reliable in pragmatic application, as well as time and cost saving.

In the future, we will extend this study from several aspects. One of the extensions is analyzing the relationships between sparsity of dataset and accuracy of adjusted LLE and NB under different application settings. Another extension is integrating this method with data mining techniques for two purposes, one is digging customer latent and affective needs to generate novel product configurations, the other one is improving design efficiency by using big data to achieve market advantages.

Acknowledgements
This study was supported by the National Natural Science Foundation of China under Grant No. 71571023. The authors are grateful to appreciate the reviewers' comments and effort in order to improve the paper.

References

[1] Wang Y and Tseng M M 2015 *J. Intell. Manuf.* 26 501-9
[2] Kusiak A 2006 Int. J. Pro. Res. 44 4175-91
[3] Zhou F Ji Y and Jiao R J 2012 J. Intell. Manuf. 24 1047-69
[4] Shepperd M and Cartwright M 2011 IEEE T. Software Eng. 27 987-98
[5] Stumptner M 1997 AI Commun. 10 111-25
[6] Wang Y and Tseng M M 2011 AI EDAM 25 185-95
[7] Wang Y and Tseng M M 2011 CIRP Ann-M. 60 175-8
[8] Keogh E and Mueen A 2011 Curse of Dimensionality ed C Sammut and G I Webb (New York: Springer) pp 257-8
[9] Xie H Henderson P and Kernahan M 2005 Int. J. Pro. Res. 43 4455-69
[10] Felfernig A Hotz L Bagley C and Tiihonen J 2014 Knowledge-based Configuration: From Research to Business Cases (Waltham, MA: MK Publications)
[11] Haug A Hvam L and Mortensen N H 2012 Comput. Ind. 63 471-81
[12] Felfernig A Schubert M and Zehentner C AI EDAM 26 53-62
[13] Nagamachi M and Lokman A M 2011 Kansei/Affective Engineering (Boca Raton, FL: CRC Press)
[14] Nagamachi M 2002 Appl. Ergon. 33 289-94
[15] Tseng M M Jiao R J and Wang C 2010 CIRP Ann-M. 59 175-8
[16] Mattelmäki T and Battarbee K 2002 Proc. 7th Biennial Paricipatory Design Conf. June 23-5 Malmø, Sweden
[17] Lin J and Seepersad C C 2007 Proc. 19th Int. Conf. Design Theory and Methodology September 4-7 Las Vegas
[18] Leonard D and Rayport J F 1997 Harvard Bus. Rev. 75 102-13
[19] Hughes R T 1996 Inform. Software Tech. 38 67-75
[20] Aspinall W P and Cooke R M 2013 Risk and Uncertainty Assessment for Natural Hazards ed J Rougier S Sparks and L J Hill (Cambridge: Cambridge University Press) pp 64-99
[21] Eggstaff J W Mazzuchi T A and Sarkani S 2014 Reliab. Eng. 121 72-82
[22] Roweis S T and Saul L K 2000 Science 290 2323-6
[23] Wan X Wang D Tse P W Xu G and Zhang Q 2016 Measurement 78 138-50
[24] Russell S J and Norvig P 2009 Artificial intelligence: A modern approach (Upper Saddle River, NJ: Prentice Hall)
[25] Özgür A Özgür L Güngör L 2005 Text Categorization with Class-based and Corpus-based keyword selection ed P Yolum T Güngör F Gürgen and C Özturan (Berlin: Springer-Verlag) pp 606-15
[26] Daubechies I, Defrise M and De Mol C 2004 Com. PA Math. 57 1413-57

IOP Conf. Series: Materials Science and Engineering **261** (2017) 012023 doi:10.1088/1757-899X/261/1/012023

Sentiments Analysis of Reviews Based on ARCNN Model

Xiaoyu Xu[1], Ming Xu[1], Jian Xu[1], Ning Zheng[1] and Tao Yang[2, *]

[1]School of Computer Science and Technology, Hangzhou Dianzi University, Hangzhou, China
[2]Key Lab of the Third Research Institute of the Ministry of Public Security

[*]yangtao@stars.org.cn

Abstract. The sentiments analysis of product reviews is designed to help customers understand the status of the product. The traditional method of sentiments analysis relies on the input of a fixed feature vector which is performance bottleneck of the basic codec architecture. In this paper, we propose an attention mechanism with BRNN-CNN model, referring to as ARCNN model. In order to have a good analysis of the semantic relations between words and solves the problem of dimension disaster, we use the GloVe algorithm to train the vector representations for words. Then, ARCNN model is proposed to deal with the problem of deep features training. Specifically, BRNN model is proposed to investigate non-fixed-length vectors and keep time series information perfectly and CNN can study more connection of deep semantic links. Moreover, the attention mechanism can automatically learn from the data and optimize the allocation of weights. Finally, a softmax classifier is designed to complete the sentiment classification of reviews. Experiments show that the proposed method can improve the accuracy of sentiment classification compared with benchmark methods.

1. Introduction

Since entering the 21st century, the Internet has gradually become an indispensable part of people's lives. With more and more people join the network, a large number of network information will be generated. This information contains both the objective facts that already exist and the subjective emotions, and the text is one of the most important ways in which the information presents to people.

Sentiment analysis is an important research in the natural language processing filed. To do this research, the first step is to convert the text and feature into a digital sequence. It is common to use bag of words or n-grams to represent text [1]. However, these methods are not only has a high-dimensional representation of word, but ignore the whole words order in the text, which limits the functionality of the learning models based on such representation. In recent years, the distributed representation methods have brought new solutions to learn better representation of text. The most commonly used methods are Word2Vec [2] and GloVe [3] that have a good analysis of the semantic relations between words and solves the problem of dimension disaster.

Based on these represent, the earliest studies of sentiment analysis focus on machine learning based methods [4, 5] and lexicon based method. The performance of machine learning methods such as Support Vector Machine [5], Naïve Bayes, Maximum Entropy model, Random Walk model strongly depends on the quality of the extracted features, and the performance of lexicon based method relied heavily on the quality of emotion lexicon [6]. These methods can learn the shallow features of a single word accurately, but cannot learn the semantic relations between words and words in the text. In addition, many researchers also use deep learning methods to deal with the sentiments analysis, such as the convolution neural network (CNN) [7] and recurrent neural networks (RNN) [8]. CNN treats each words fairly and uses sliding windows which also called filters to do the feature mapping, then a pooling is to get the fixed length output. Unlike the CNN, RNN takes into account the order of words and could better deal with the text. Although some models based on the deep neural network can learn the deep semantic relation between the word and the word and the sentence and the sentence, it does not highlight the characteristics of the single word.

In view of the above problems, this paper mainly uses the deep learning methods to study the sentiment analysis of the reviews. We build the bidirectional recurrent neural network to handle the input with variable comment text. And inspired by recent success of attention mechanism [9], we combine the attention mechanism with BRNN to focus on the specific scope of the text. This is important for the semantic extraction of text and its visibility can directly explain the importance of each word to the sentiment analysis. Then, we add a convolution process which can better handle local information. We have evaluated our proposed method with benchmark experiments. Results reveal in Section 4 and show that it is effective to use the ARCNN to analyse the sentiment.

The rest parts of this paper are organized as follow. The following section reviews the related work. Next, the method for sentiment analysis is described in Section 3. And Section 4 reports the results of our experiments. Finally, Section 5 draws a conclusion.

2. Related work

In recent years, the neural network based sequence model has been a great success in text processing, especially in comparison with the BOW (bag of words) model, showing significant advantages. CNN has made a lot of outstanding achievements in many aspects, such a parsing, search word retrieval, sentence model and other traditional natural language work. CNN uses the local spatial correlation between layer and layer, and the neurons of each adjacent layer is only connected with the closer previous neuron node, which greatly reduces the architecture parameters size of the Neural Network [10]. In addition to CNN, RNN also been used in the NLP [11, 12]. In the field of machine translation, the introduction of RNN makes the dialogue model have made new progress. Sutskever et al. [13] proposed an end-to-end sequence learning method that minimizes the computational complexity of sequence structures. In the task of sentiment analysis, both of CNN and RNN can play a good role. Kim et al. [7] used a simple convolution neural network and word vectors as the input of model. Liang et al. [8] extended LSTM to recursive neural network to capture the deeper level of semantic syntax information. Wen et al. [14] proposed a RCNN with Highway network to learning text representation for sentiment analysis.

And since Denil et al. [15] proposed attention mechanism, the deep learning model which relies on this mechanism has been successfully applied to a variety of NLP tasks. In 2014, Bahdanau et al. [16] used the attention model to adjust the input and output sequences of machine translations and achieved good results. In 2015, Xu et al. [17] proposed an attention-based model to automatically learn to describe the contents of the images. In the same year, Sharma et al. [18] divided the attention model into two categories, soft attention and hard attention, and the method can be used to learn the weight of attention.

3. Method

The figure 1 gives architecture of our method. Next we introduce it in several major modules.

3.1. Input

For the traditional word representation, such as one-hot representation [1], the biggest problem is that it can't analyse the semantic relation between words, and its dimension is particularly large. Recent research has demonstrated distributed representation [2, 3] of word embedding is more powerful. A word can be represented by a vector as follows:

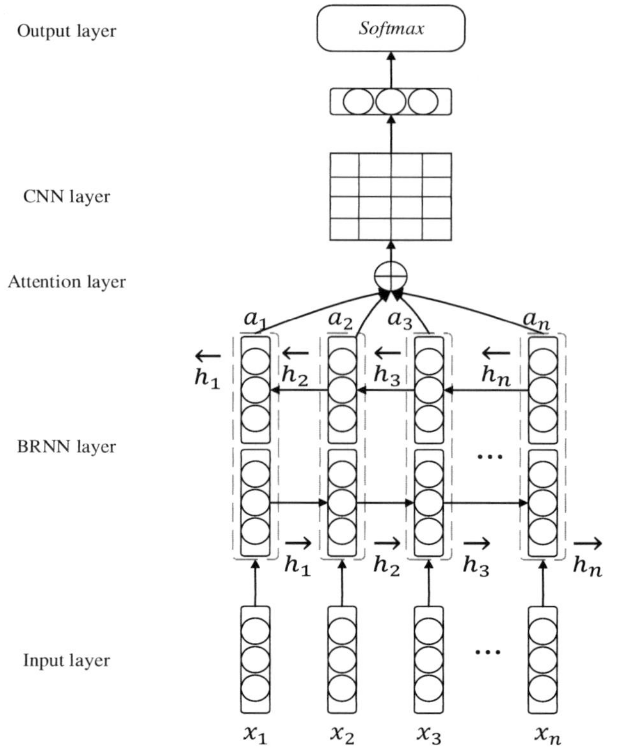

Figure 1. The architecture of the proposed model

$$W = Lx \tag{1}$$

where $x \in R^{|N|}$ is a one-hot vector and $L \in R^{d \times |N|}$ is a word-representation matrix. In fact, it is a mapping from high dimension to low dimension. In this paper, we use *Glove* embedding proposed by [3]. The model is trained by minimize the loss function:

$$J(\theta) = \frac{1}{2}\sum_{i,j=1}^{W} f(P_{ij})(u_i^T v_j - \log P_{ij}) \tag{2}$$

3.2. BRNN and Attention mechanism

The recurrent neural network is a special feedforward neural network, which is very effective for modelling a time-dependent sequence. The characteristics of its loop structure make the output of each moment associated with the output of the previous time, which is very suitable for dealing with variable-length text, and can well express the correlation between nodes in the sequence. Most of the existing models compile the input sequence into a fixed-length vector, which ignores the characteristics of the dimension that is important for the expression of the sentence and severely limits the resulting semantic vector. The advantage of the recurrent neural network is that its recurrent structure can handle non-fixed-length vectors and can keep the time series information.

The single RNN derives the characteristics of the current word based on the output characteristics of the previous word, which takes into account only one side of the text and does not take into account the effect of the next word on the current word. In order to make full use of all the information contained in the text sequence, this paper uses the Bidirectional Recurrent Neural Networks (BRNN) [19] to encode the input sequence. The network in parallel two time directions, and at every moment, the past information and feature information can be used directly. The hidden layer state is calculates as follows:

$$\overrightarrow{h_t} = f(\vec{U}x_T + \vec{V}h_{t-1} + \vec{b}) \quad (3)$$
$$\overleftarrow{h_t} = f(\overleftarrow{U}x_T + \overleftarrow{V}h_{t+1} + \overleftarrow{b}) \quad (4)$$

where the f is an activation function, b is the offset and U, V are the shared weight parameters of the entire neural network. In addition, $\overrightarrow{h_t}, \overleftarrow{h_t}$ represent the forward hidden state and backward hidden state respectively. We represent the final state at position t with the concatenation of $\overrightarrow{h_t}$ and $\overleftarrow{h_t}$.

$$h_t = \overrightarrow{h_t} \parallel \overleftarrow{h_t} \quad (5)$$

After the BRNN, we consider the attention mechanism. The central idea of the attention mechanism is to let the model automatically find information about the relevant parts (usually as vector sequences) from the stored memory and use the information to get the target results. The advantage of using the attention mechanism is that, firstly, a complete review usually consists of many factors, and the model can learn how to effectively integrate the various factors into a whole, rather than simply translating information from one mode to another mode.

In this part, we use a 2-layer feedforward neural network [16] to train attention scores and the number of memory cells is equal to the BRNN hidden unit. The attention mechanism gives an attention score $e_{i,t}$ to each word t in the sentence i.

$$e_{i,t} = \mathrm{g}(Wh_t, \mathrm{c}) \quad (6)$$

where the g is an activation function. And then, the weight probability $a_{i,t}$ of each h_t is computed by

$$a_{i,t} = \frac{\exp(e_{i,t})}{\sum_{k=1}^{T} \exp(e_{i,k})} \quad (7)$$

3.3. CNN and Output

Convolutional layers can encode significant information about input data with significantly fewer parameters than other deep learning. And as mentioned in the previous section, we use the attention model to connect CNN and BRNN to obtain the better feature representations.

For the CNN, given a window size w, a filter is seen as a weight matrix $f = (f_1, f_2, \ldots, f_w)$ (f_i is a column vector), and the bias is seen as b. A feature map can be obtained by an activation function f. Each output map could be a value combining by multiple filters, and it can be calculate as follow:

$$z_i = f(\sum_{j=0}^{w-1} f_{j+1}u_{j+i} + b) \quad (8)$$

where z_i and u_i is an element in the output and input, respectively. And we use the max pooling to further abstract the features generated from the convolutional layer.

The output of the above layers is regarded as features of sentences for sentiment classification. At the end of the model, there is a full-connection *softmax* classifier which gets the output of each category. We use the e $p(i|x, \theta)$ to denote the probability of the review being class i, where θ is the parameters of model:

$$p(i|x, \theta) = \frac{exp(y_i)}{\sum_{j=1}^{k} exp(y_j)} \quad (9)$$

4. Experiments

4.1. Dataset

In this section, experiments are conducted to evaluate the performance of the proposed algorithm. The dataset used in the experiments comes from the Kaggle competition, and it contains 25000 movie reviews. Each review is labelled by human with 1 represents positive sentiment and 0 represents negative sentiment. Data sample showed as table 1. We randomly selected 5000 reviews as the test set, leaving the 20000 as the training set.

Table 1. Data sample

Positive	Tt is a fantastic film! I've not laughed so much when watching a film for ages!
	This is a wonderful story of a woman's bravery, courage and extreme loyalty.
Negative	This is an awful film. Yea the girls are pretty but it's not very good.
	I do not recommend this movie , because it's inaccurate and misleading

4.2. Results and evaluation

In this section, we designed three experiments:

- BRNN: only use the Bidirectional Recurrent Neural Network to do sentiments analysis;
- ARNN: combine the attention mechanism with BRNN to do sentiments analysis;
- ARCNN: combine the attention mechanism with BRNN-CNN to do sentiments analysis.

We use the GloVe algorithm to train the vector representations for words and adopt softmax output to make prediction. The word embedding is pre-trained in a corpus of about 400000 in size, and the dimension of word vector is set to 100. For our model, ARCNN, we use 'RMSprop' [20] as training method with batch size of 50, the BRNN hidden layer dimension of 100, and the CNN filters number of 64.

In order to evaluate the effectiveness of our method for sentiment analysis of reviews, we use the accuracy as evaluate metric. The accuracy reflects the ability of the classifier system to judge the whole sample and is defined as:

$$accuracy = \frac{TP + TN}{TP + FP + TN + FN} \quad (10)$$

where the *TP* is the number of reviews that are correctly divided into positive review, i.e. the number of the reviews that are classified as positive and its real category is positive too. The *TN* is the number of reviews that belong to negative category and assigned with negative. The *FN* is the number of reviews that belong to positive category and assigned with negative. And the *FP* is the number of reviews that belong to negative category and assigned with positive.

The figure 2 shows the accuracy of three methods in the training phase and figure 3 shows the accuracy in the test phase. We can observe from these figures that the accuracy of the ARNN and ARCNN are superior to the BRNN method. We take the best results of 10 iterations and show them in the table 2. By comparing the test accuracy of BRNN (80.22%) with ARNN (89.98%), it is found that using the attention mechanism with the BRNN leading to higher accuracy than only using the BRNN model. The reason is that the BRNN model cannot specifically capture the sentence information and does not take into account the different priorities of each word in the sentence. And the ARNN model with the attention mechanism can automatically assign low weights to irrelevant words, and high weights are assigned to the relevant words that express emotions. In addition, because of the advantage of CNN for the local information processing, the accuracy of ARCNN (90.34%) is slightly better than ARNN (89.98%).

Table 2. Three methods comparing

Method	Stage	Accuracy (%)
BRNN	Training	89.50
	Test	80.22
ARNN	Training	95.55
	Test	89.98
ARCNN	Training	**96.64**
	Test	**90.34**

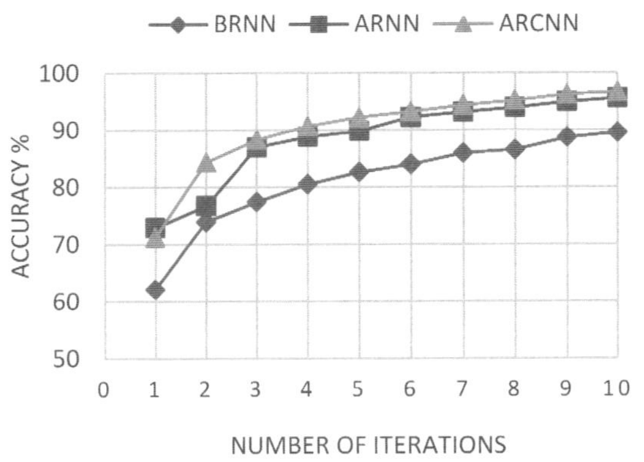

Figure 2. The accuracy of the training stage.

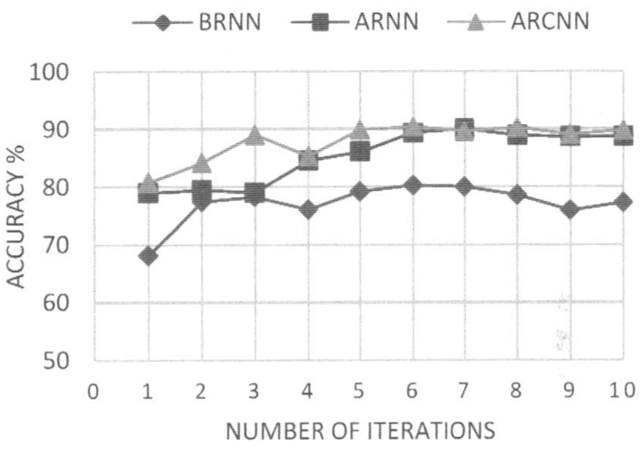

Figure 3. The accuracy of the test stage.

5. Conclusion

In this paper, we propose a sentiment classification method, named ARCNN, which makes the combination of attention mechanism and BRNN-CNN. In the model, the BRNN can review the information of each time series entered before, and add the attention mechanism to optimize the weight allocation in its hidden layer output. Then enter the convolution neural network to get better features. The design of the model takes into account the calculation of attention weight and the mining of information. The experiments show that the model proposed in this paper can capture the context and assign the context-related weights to each word, so as to effectively excavate the important words and improve the accuracy of classification.

Acknowledgments

This work is supported by the National Key R&D Plan of China under grant no. 2016YFB0800201, the Natural Science Foundation of China under grant no.61070212 and 61572165, the State Key Program of Zhejiang Province Natural Science Foundation of China under grant no. LZ15F020003, the Key research and development plan project of Zhejiang Province under grant no. 2017C01065, the Key Lab of Information Network Security, Ministry of Public Security, under grant no C16603.

References

[1] Murphey Y L, Masrur M A, Chen Z, et al. Model-based fault diagnosis in electric drives using machine learning [J]. IEEE/ASME Transactions on Mechatronics, 2006, 11(3):290~303.

[2] Mikolov T, Sutskever I, Chen K, et al. Distributed representations of words and phrases and their compositionality[C]. Advances in neural information processing systems. 2013: 3111-3119.

[3] Pennington J, Socher R, Manning C D. Glove: Global Vectors for Word Representation [C]. EMNLP. 2014, 14: 1532~1543.

[4] Go A, Bhayani R, Huang L. Twitter sentiment classification using distant supervision [J]. CS224N Project Report, Stanford, 2009, 1(12).
[5] Pang B, Lee L, Vaithyanathan S. Thumbs up?: sentiment classification using machine learning techniques [C]. Proceedings of the ACL-02 conference on Empirical methods in natural language processing-Volume 10. Association for Computational Linguistics, 2002: 79~86.
[6] Hu M, Liu B. Opinion Feature Extraction Using Class Sequential Rules [C]. AAAI Spring Symposium: Computational Approaches to Analyzing Weblogs. 2006: 61~66.
[7] Kim Y. Convolutional neural networks for sentence classification [J]. arXiv preprint arXiv:1408.5882, 2014.
[8] Liang and Y. Cai, Polarity Shifting and LSTM Based Recursive Networks for Sentiment Analysis [J]. Journal of Chinese Information Processing, 2015. 29(5): p. 152~159.
[9] Denil M, Bazzani L, Larochelle H, et al. Learning where to attend with deep architectures for image tracking [J]. Neural Computation, 2011, 24(8):2151~2184.
[10] Clarke D. A context-theoretic framework for compositionality in distributional semantics [J]. Computational Linguistics, 2012, 38(1): 41~71.
[11] Irsoy O, Cardie C. Opinion Mining with Deep Recurrent Neural Networks[C]. Conference on Empirical Methods in Natural Language Processing. 2014:720~728.
[12] Serban I V, Sordoni A, Bengio Y, et al. Building end-to-end dialogue systems using generative hierarchical neural network models[C]. Thirtieth AAAI Conference on Artificial Intelligence. 2016.
[13] Sutskever I,Vinyals O, Le Q V. Sequence to sequence learning with neural networks [C]. Advances in neural information processing systems. 2014: 3104~3112.
[14] Wen Y, Zhang W, Luo R, et al. Learning text representation using recurrent convolutional neural network with highway layers [J]. arXiv preprint arXiv:1606.06905, 2016.
[15] Denil M, Bazzani L, Larochelle H, et al. Learning where to attend with deep architectures for image tracking [J]. Neural Computation, 2011, 24(8):2151~2184.
[16] Bahdanau D, Cho K, Bengio Y. Neural machine translation by jointly learning to align and translate [J]. arXiv preprint arXiv:1409.0473, 2014.
[17] Xu K, Ba J, Kiros R, et al. Show, attend and tell: Neural image caption generation with visual attention [C]. International Conference on Machine Learning. 2015: 2048~2057.
[18] Sharma S, Kiros R, Salakhutdinov R. Action recognition using visual attention [J]. arXiv preprint arXiv:1511.04119, 2015.
[19] Pollastri G, Mclysaght A. Porter: a new, accurate server for protein secondary structure prediction [J]. Bioinformatics, 2004, 21(8): 1719~1720.
[20] D. Kingma and J. Ba. Adam: A method for stochastic optimization. arXiv preprint arXiv: 1412.6980, 2014.